Electronic power supply handbook

UNIV STRATH

Electronic power supply handbook

Ian R Sinclair

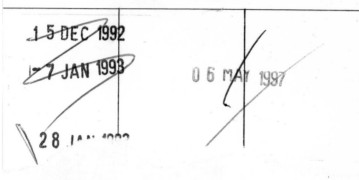

PC Publishing

PC Publishing
4 Brook Street
Tonbridge
Kent TN9 2PJ

First published 1990

© PC Publishing 1990

ISBN 1 870775 03 1

British Library Cataloguing in Publication Data

Sinclair, Ian R. (Ian Robertson)
   Electronic power supply handbook.
   1. Electronic equipment. Power supply devices
   I. Title
   621.381'044

   ISBN 1-870775-03-1

Phototypesetting by Scribe Design, Gillingham, Kent
Printed and bound by Dotesios

Preface

Power supplies are the common factor in all electronics equipment, but sufficient attention is not always paid to this type of circuit. Even the basic rules about rectification and smoothing are not always made clear in electronics courses, and many people experience considerable difficulty in extracting the information that they need from texts that always seem to relegate power supplies to a final brief chapter, and to ignore some forms of power supply altogether.

This book deals with all forms of power supplies for electronics use, including some that are considered to be exotic at the time of writing but whose importance will increase over the years. The fundamentals of batteries are strongly emphasised because so many electronic circuits are nowadays battery powered, but the rapid development of battery technology over the past few years has not been matched by a corresponding flow of information.

No apology is made for dealing with the topic of rectification and smoothing from the foundations, because this topic is all too often glossed over, and the relationship between steady current, reservoir capacitor size and ripple is seldom explained. Power supply design is not simply a matter of using a few formulae — the understanding of the processes is important if expensive mistakes are to be avoided, particularly since some of the formulae that are used are of dubious validity.

The subject of stabilisation is also thoroughly covered, less from the point of view of designing stabilisers from scratch than from the intelligent use of the many IC stabilisers which are now available. Once again, though, understanding the principles leads to much better appreciation of how to specify components correctly.

I would like to thank many who contributed to this book in various ways, by the provision of information, advice and constructive criticism. In particular I would like to acknowledge the help of RS Components, whose many publications are always of immense assistance to anyone with an interest in electronic components. I would like also to thank Phil Chapman, whose continual encouragement has made this and other books possible.

Ian R Sinclair

Contents

6 Other supplies

1 Fundamentals and components

Electric current in a circuit is predominantly a flow of electrons, though hole movement contributes noticeably in some types of materials. The materials that we call metals contain vast numbers of electrons that are not tightly bound to their atoms and so are free to move from atom to atom, the movement that we call electric current. When this movement is predominantly in one direction we call the current DC, the abbreviation of the old phrase *direct current*. When the movement is oscillatory, vibration about a set position with no overall drift in one direction, the current is AC, *alternating current*. In many cases the motion is a combination of both, with some oscillation and some overall movement, the type of current that is called unidirectional.

The idea that we can have pure DC is fictitious, because the electrons in any material are always oscillating. This is the form of AC that we call noise. In addition, there is much misunderstanding about the speed of electrons in materials. A good conductor such as copper contains a vast number of mobile electrons, so many that to carry a current of one ampere or so requires a movement that is of the order of a few centimetres per hour. Do not confuse the speed of the electrons with the speed of electricity, which is around 200,000,000 metres per second. A wire is packed with electrons, and when one electron moves in, one will move out at the other end with almost no delay; the speed for this process is the speed that we measure for electric current. The fact that each individual electron moves slowly is not a contradiction. Think, for example of a hosepipe filled with water. When the water is turned on at the tap, the water in the pipe might move fairly slowly, but there is an instant response at the other end if the hose was indeed filled.

Electron movement is considerably faster in some other materials. When electrons move in a vacuum, the speed can be greater than 200,000,000 metres per second, and can approach the limit of

300,000,000 metres per second, the speed of light. This is possible because in a vacuum there is nothing to impede the movement of the electrons, and because there are very few electrons available their speed has to be very high to carry a current of even a few microamperes. The speed of electrons within semiconductors is also higher than in copper or other good conductors, again because the number of electrons is not so great. In addition, some of the current in a semiconductor will be carried by hole movement in the direction opposite to the movement of the electrons.

Steady-voltage supplies

By far the majority of electronic circuits require a power supply which is a steady DC voltage, certainly not AC, and in most cases containing no AC or only the smallest trace of AC. This DC can be obtained in three main ways, from electromechanical generators (dynamos), from chemical or other energy conversions (batteries) or from AC mains by rectification and filtering. Of all these sources, batteries provide the smoothest DC output, because the output from any chemical converter is naturally smooth DC. Mechanical generators which depend on commutators and brush-gear provide an output which requires considerable smoothing, and even the homopolar generator, whose output is, like that of a battery, naturally DC, requires smoothing because of the irregularities of brush contact. The DC that is obtained by rectification is really only unidirectional, even if a polyphase supply is used, and the action of a reservoir capacitor is essential if the output is even to resemble DC.

A supply that is not of perfectly smooth DC is said to have *ripple*, and the ripple can be measured as current or as voltage. The ripple current or voltage can be quoted in terms of peak-to-peak value, or it can be assumed to be a sinewave and its RMS value quoted. For a sinewave, the RMS value is 0.707 times the AC peak value or 0.3535 times the AC peak-to-peak value, Figure 1.1. The ripple waveform is most unlikely to be a sinewave; a much more usual form is a sawtooth or triangular wave, but since the *comparative* ripple amplitude is the factor that is of interest, the RMS value is more often quoted. The peak-to-peak value, which is more likely to be the measured value on an oscilloscope, is then found by multiplying the RMS value by 2.8.

Ripple is never a constant, and it is misleading to quote ripple amplitude without also quoting the current load at which the ripple is measured. Many power supplies have their ripple

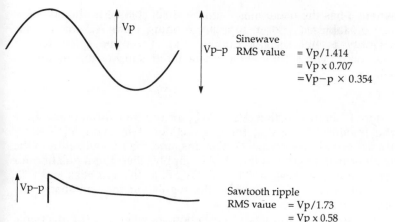

Sinewave
RMS value $= Vp/1.414$
$= Vp \times 0.707$
$= Vp\text{-}p \times 0.354$

Sawtooth ripple
RMS value $= Vp/1.73$
$= Vp \times 0.58$
$= Vp\text{-}p \times 0.29$

Figure 1.1 Peak, peak-to-peak and RMS relationships for a sinewave and for a sawtooth. For calculations of RMS ripple, the factor of 0.35 is normally used for conversion from peak-to-peak to RMS value, though this is only an approximation.

amplitude quoted at low current outputs, though it is now becoming more common for manufacturers to show the amount of ripple when the supply is operating at full rated current, a much more useful measurement. The amplitude of ripple can be expected to increase considerably if the rated current output is exceeded, whether the supply is electronically stabilised or not. Stabilisation is not a cure-all for power supplies, and when the current rating of a stabilised supply is exceeded, the voltage drop and the amount of ripple can both be very large.

Regulation and internal resistance

For any form of DC power supply, the output voltage will change as the line (AC) voltage changes and as the load current changes. The regulation of the supply measures how well it can cope with such changes. For line voltage changes, the line regulation can be quoted as the change in output voltage per change in AC input voltage:

$$\frac{\delta V_{out}}{\delta V_{in}}$$

where δ has the usual meaning of a small change in the quantity. For unstabilised power supplies running from AC mains, this quantity is not usually quoted, but it can be calculated or measured. For a completely unstabilised supply, the regulation will be:

F1 × F2

where F1 is the fraction AC_{out}/AC_{in} for the transformer, and F2 is the fraction DC_{out}/AC_{in} for the rectifier. For example, if a 9V transformer is used from 240V mains, and the DC output from the rectifier is 0.8 (80%) of the 9V AC supply, then the regulation for line changes is 9/240 × 0.8 = 0.03V. For 110V supplies and 80% DC output of the 9V supply, the regulation for line changes is 9/110 × 0.8 = 0.065V.

A more usual and useful definition, however, is the fractional or percentage change in output divided by fractional or percentage change in input. For an unstabilised supply, the output will, other factors being equal, always be the same fraction of the AC line input, so that this measure of regulation gives the result of unity. For a stabilised supply, the figure should be considerably lower.

The effect of load current, when we consider a completely unstabilised supply, is that the output voltage will drop as the output current increases. The less the drop in output voltage, the better the output regulation of the power supply is said to be, and the ratio:

$$\frac{\delta V}{\delta I}\text{(change of voltage divided by change of current)}$$

is called the internal resistance of the power supply and should be as low as possible for good stabilisation.

For example, if the output of a nominal 9V supply changes to 8.85V when 120mA of current is drawn, then the internal resistance is:

$$\frac{0.15}{0.12} = 1.25 \ \Omega$$

which is quite a small amount for a mains supply.

A stabilised supply can be designed so that over a limited range of voltage and current the output voltage will rise slightly when the output current increases. This corresponds to the internal resistance having a negative value. This is normally an unstable situation which will lead to oscillation, and it is used only to compensate for other resistances that are present. Most stabiliser

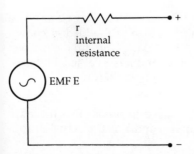

Figure 1.2 The simple equivalent circuit of a power supply of any kind consists of a source of steady voltage E and a fixed series resistance r, the internal resistance.

ICs, however, feature very large internal gain and are liable to oscillation unless bypassed with a suitable capacitor — more details are noted in Chapter 4.

The idea of internal resistance means that a power supply can be represented by the simple equivalent circuit of Figure 1.2, consisting of a constant-voltage supply in series with a resistance. The value of constant voltage will be the open-circuit (no-load) voltage of the supply, often called the *EMF*. The resistance in the equivalent circuit is the internal resistance of the supply, and this will include such factors as transformer winding resistances, diode drops and rectifier efficiency for a mains supply. For a battery supply, the internal resistance is the resistance of the chemicals that form the electrolyte of the battery.

The simple equivalent circuit is, in fact, usable over a reasonable range of load currents, but it has limitations at the extremes of load. The assumption of constant voltage for the generator cannot be sustained if the open-circuit voltage alters — this will happen, for example, when a battery has suffered a momentary overload. The assumption of a constant internal resistance is also invalid because, although some of the resistances that contribute to internal resistance are fixed, such as transformer winding resistance, others are not, like diode resistance, and some of the internal resistance will measure effects like reservoir capacitor efficiency for an AC powerpack.

Using the simple equivalent

The simple equivalent, in spite of its limitations, functions well for a power supply of any type that is working well within its limits.

The next point to look at is how this equivalent can be used to predict the effect of a load, whether the size of the load is known as a current or as a resistor value. Taking the first example, suppose that the load current is known, and has a value I amperes. The output voltage from the power supply is therefore:

$V = E - rI$

where r is the value of internal resistance in ohms, E is the EMF (no-load output voltage) for the supply, and V is the actual output voltage.

In a practical case, we might be using a supply whose open-circuit output voltage was 13V and whose internal resistance was 1.2 Ω. For a 1.5A current drawn by the load, the actual delivered voltage would be:

$$13 - 1.2 \times 1.5 = 11.2V$$

assuming as usual that the load current of 1.5 A is well within the rated limit for the power supply.

The short-circuit current for the power supply cannot precisely be calculated from knowing the internal resistance, but the calculation is a useful guide. In this example, the predicted short-circuit current is E/r, which gives $13/1.2 = 10.83A$. This is an overestimate, because neither the zero-load voltage nor the value of internal resistance can be expected to remain constant under short-circuit conditions. Nevertheless, the estimate is useful because it shows the margin of the power supply, the difference between the normal limit of current supply and the absolute theoretical limit. Any fuses used in the circuit should blow at a current less than this theoretical limit, otherwise there is no real protection offered by the fuse (see later in this chapter).

When the load resistance is known, the equivalent circuit is as shown in Figure 1.3, in which the load resistance is R. The output voltage is now:

$$V = E \times \frac{R}{R + r}$$

where the symbols E, V and r have the same meaning as before. For example, if a power supply has a no-load voltage of 15V, an internal resistance of 1.8Ω, and feeds a load of 10Ω, then the output voltage should be:

$$V = 15 \times 10/11.8 = 12.71V$$

making the usual assumptions.

These examples are for unstabilised power supplies, and are, if

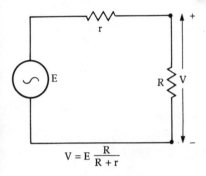

$$V = E \frac{R}{R + r}$$

Figure 1.3 Using the simple equivalent circuit to show the effect of connecting a load to the power supply. This assumes that both E and r are constant quantities.

anything, rather on the optimistic side as regards internal resistance. The effect of diode voltage drop is not very well taken account of in the simple equivalent circuit, and since the usual figure for internal resistance includes the forward resistance of the diode (not incorporating the drop) a useful way of allowing for diode drop is to measure the no-load voltage and use this figure, or if using a calculated figure, subtract whatever diode drop is applicable from the no-load voltage. Tables for rectifier circuits (Chapter 3) usually ignore the diode drop, which can be 0.6V or 1.2V, depending on whether one or two diodes are in the current path.

Electromechanical generators

When the discovery was first made by Michael Faraday in 1831 that electric current could be generated by moving a conductor through the field of a magnet, the form of generator that was used provided DC. If it had generated AC, it is doubtful if the effect would have been noticed, because all current detectors at that time were DC only, AC had not been discovered.

Faraday's generator is called a homopolar generator, and its principles are illustrated in Figure 1.4. The copper disc is placed so that it is crossed by the field of a powerful magnet, and brush contacts are arranged to touch the disk, one at the rim and the other at the shaft. As the disk spins, a voltage is induced between rim and shaft, and this is pure DC apart from the inevitable fluctuations caused by the irregular contact of the brushes. The Faraday apparatus generated just enough voltage and current to

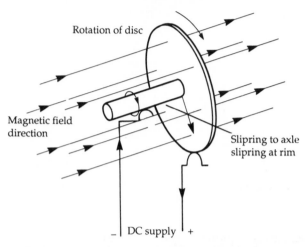

Rotation of disc

Magnetic field
direction

Slipring to axle
slipring at rim

_ | DC supply | +

Figure 1.4 Principles of the homopolar generator - the first type of generator invented by Faraday. This device generates DC, but at a very low voltage.

operate a sensitive galvanometer, but was not considered useful as a way of generating electrical power for any practical purposes.

The advantages of the homopolar generator are considerable, however. It is quiet, its output can be very smooth and though the voltage is generally low, the current can be high. The disk has to be spun at a high speed in a very strong field to achieve a useful output, but with modern permanent magnets and the high speeds that can be achieved with a gas turbine driving the disc, respectable voltages along with large current capabilities are possible, and the homopolar generator is returning as a way of powering semiconductor equipment. One advantage for military power supplies is that the homopolar generator generates only negligible amounts of RF interference.

The bulk of electromechanical generation, however, makes use of dynamos or alternators. The difference between these two is one of detail — the dynamo makes use of a mechanical form of rectification called a commutator, and the low cost of diode rectification is now making the dynamo a rare sight, since an alternator and diode circuit can perform much better. The basic form of either device is of a coil which rotates in a magnetic field, with connections to the rotating coil being made by way of brushes bearing either against rings (called slip-rings) for an alternator or against the segments of a commutator for a DC generator or dynamo.

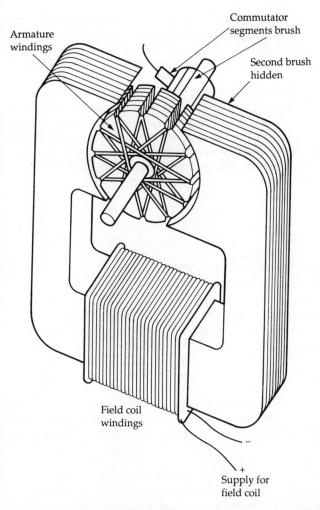

Armature
windings

Commutator
segments brush

Second brush
hidden

Field coil
windings

−

+

Supply for
field coil

Figure 1.5 Outline of a dynamo. The use of many turns of wire greatly increases
the EMF, but the rotation of the wire makes the form of the voltage an AC wave.
This is rectified by using a commutator to switch the connections at each point in
the rotation where the EMF is zero.

A typical dynamo is illustrated in Figure 1.5. The magnetic field
is provided in this example by a winding over a laminated steel
core. The steel has enough remanence (residual magnetism) to
provide a weak field when no power is being generated, and when
the dynamo is powered the voltage that is generated will pass

current through the field winding and so greatly increase the strength of the magnetic field, so in turn increasing the generated voltage. The output voltage varies with the speed of rotation, and this variation can be almost linear with rotational speed, a fact used for tacho-generators.

The variation of dynamo output with speed makes it necessary either to run a dynamo at a constant speed, using mechanical regulators, or to use some form of electrical constant-voltage regulation. Since the output of a dynamo at any rotational speed is proportional to the strength of the magnetic field, the output can be regulated by altering the current passing through the field winding. By reducing this field current, the dynamo output will be reduced and vice versa, so that the basis of control of dynamo output is the use of negative feedback to the field winding. As the field winding is normally fed from the dynamo output, constituting positive feedback, quite large changes in the field current are needed to restore control, and at very high rotational speeds the residual magnetism of the field core metal can be enough to supply the required voltage output with no field current flowing.

The methods used in the past, notably on dynamos in cars, were all electromechanical, based on a contact-breaker system which would switch over either on excessive current or on excessive voltage. This was done by using two coils on a relay core, one of a few turns which carried the output current and one of many turns which was connected between + and − outputs to sense voltage. When the contacts, Figure 1.6, were opened because of excessive current or voltage, the field current was reduced to a much lower value through a series resistor so reducing the output. The lower output then caused the contacts to close again, and the action of the device was oscillatory, opening and closing several times per second with a mark/space ratio which depended on the dynamo speed. The faster the dynamo was rotated, the greater the fraction of time that the contacts remained open, reducing the field current. This caused the output voltage from the dynamo to be a low-frequency square wave which was smoothed by its connection to a lead-acid battery.

Such a method of control would be quite useless for any equipment that required smooth DC and which did not have the smoothing action of a storage battery. Dynamo generators for powering electronic equipment, when no storage battery is used, must be designed for constant output voltage, and this is usually done by controlling the speed of the driving shaft. This can be supplemented by control of the field current, but using electronic regulators rather than the older switching electromechanical type.

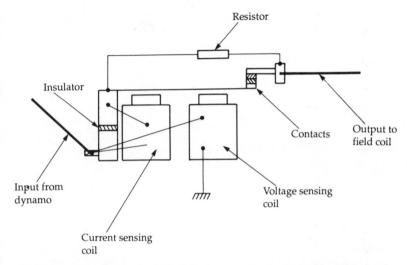

Figure 1.6 Outline of the old type of regulator system for controlling dynamo output, using both a voltage and a current sensing coil to control the current in the field winding.

If the field current is to be controlled smoothly, however, this calls for analogue methods rather than switching methods, so that high-power output transistors are needed to handle the field current, which can be large. In addition, since the field winding has a large inductance, any semiconductor controllers have to be well protected against induced voltages which occur when the field current is changed rapidly.

The output from any dynamo that uses a commutator and brushgear will be irregular, with large amplitudes of electrical noise ('hash') from the brushes because of the localised sparking that occurs there. Since many generator supplies are intended for use with radio transmitters and receivers, the RF interference generated in this way must be suppressed, and the dynamo output also needs to be smoothed to eliminate the lower-frequency fluctuations. The smoothing requirements are eased by using polyphase armature windings, but this does nothing to reduce the hash, and usually increases it. Suppression of a dynamo cannot be carried out in any simple way, and needs a combination of small ferrite-cored inductors and bypass capacitors, using circuits approved by the manufacturers of the dynamo. It is important to observe restrictions on capacitor size and positioning, because a capacitor placed incorrectly in the feedback circuit can cause the output of

11

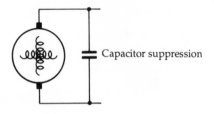

Capacitor suppression

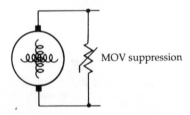

MOV suppression

Suppressors are connected across brushes

Figure 1.7 Typical dynamo-suppression methods to reduce the RF interference caused mainly by sparking at the commutator. The suppressing devices are connected across the brush terminals.

the dynamo to 'hunt', so that the output voltage rises and falls at regular intervals as the capacitor charges and discharges. Figure 1.7 shows typical recommendations for the suppression of a car dynamo in the days when cars used DC generators. Many commercial suppression circuits make use of metal-oxide varistors (MOVs), a form of voltage-dependent resistor.

The undesirable features of dynamos, combined with the reliability and low price of silicon diodes, have lead to the widespread use of alternators, particularly for cars where the easier control that is possible using alternators allows the use of alternators which can produce useful charging current even at low rotational speeds without the risk of running out of control at higher speeds.

The greatest single advantage of using an alternator for power generation is that the brushes are in contact with slip-rings (Figure 1.8) rather than commutator segments, so that no current switching is being carried out at the brushes. This enormously reduces the amount of electrical noise at the brushgear, even if it

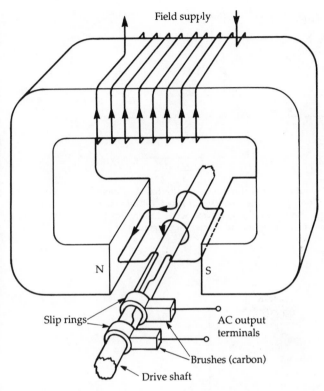

Field supply

N   S

Slip rings

AC output
terminals

Brushes (carbon)

Drive shaft

Figure 1.8 The slip-ring contacts of a typical alternator. Since there is no switching involved, sparking and the consequent interference are both much reduced.

does not entirely eliminate interference pulses. The largest single step to the widespread use of alternators, particularly in cars, has been the availability of high-current (50 A or more) silicon diodes with low forward resistance.

The layout of a typical car alternator is illustrated in Figure 1.9. The design is inside-out compared with the older alternator types in the sense that the field coils form the rotor, and are energised by way of slip-rings. Since the energising current is much lower than the normal rated output current of the alternator, this further reduces brush noise. The stator consists of the coils in which the output is induced, and these are usually arranged at 120° to each other so that the output AC is three-phase. A three-phase diode bridge, Figure 1.10, then rectifies this output, and in car applications the final smoothing is, as usual, by way of the storage

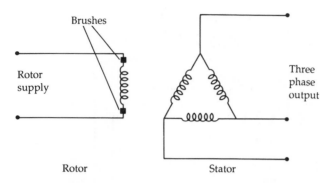

Figure 1.9 The layout of a typical modern car alternator, showing the three-phase stator coils and the single-phase rotor which is fed with its magnetising current through slip-rings. No rotating contacts are needed in the main current path from the stator, making this a much more efficient arrangement.

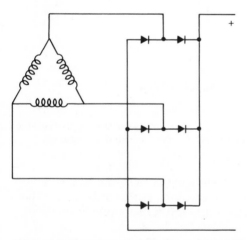

Figure 1.10 The form of a three-phase diode bridge for rectifying the output from an alternator.

battery. For car use, the only suppression that is required is a plastic-film capacitor of about 2µ2 connected between the output terminal (following the rectifier) and earth at the alternator. Note that capacitors must never be connected across the field winding of an alternator or dynamo because this could cause a delay in the control action that could lead to instability.

14

Control of the car alternator is by way of a switching transistor which operates, as usual, on the field current. The output voltage is detected in a comparator whose other input is from a Zener diode, and the output of the comparator is used to carry out the switching. This results in a high-frequency signal being superimposed on the normal AC output (which in any case is at an audio frequency of several kHz), but the rectification and smoothing of the output makes this less serious.

Alternators are available in a very large range of sizes from a few watts upwards, and the smaller types can usually be obtained along with rectifier bridges and control circuits. Any form of engine drive can be used, and one of the considerable advantages of modern alternators is that the output DC voltage from the rectifiers can be almost independent of the speed of rotation of the alternator, making this ideal for variable-speed power sources such as wind and water generators.

Electrolytic capacitors

Smoothing of DC supplies is carried out mainly by capacitors of the electrolytic type, and a good working knowledge of the principles of this type of capacitor is essential for anyone working with power supplies. In addition, large-value electrolytic capacitors of several farads are often used in place of cells for back-up purposes in digital equipment in the event of mains failure. In general terms, a capacitor is an arrangement of conductors that are insulated from each other so that charge can be stored. Imagine two metal plates arranged parallel to each other and separated by air. If we apply a voltage to the plates by connecting them to a battery, Figure 1.11, then one plate will gain electrons and the other will lose electrons. When the battery connections are removed there is no connection between the plates that would allow the electrons to return, and there will still be a voltage between the plates, caused by the fact that one plate is negatively charged and the other positively charged. This arrangement is a capacitor. If the plates are now connected, a transient current will flow until there is no surplus or deficit of electrons on either plate. The total amount of charge that flows when the plates are connected in this imaginary experiment can be calculated by multiplying the amount of current that flows by the time for which it flows, but this is not a simple calculation because the current is

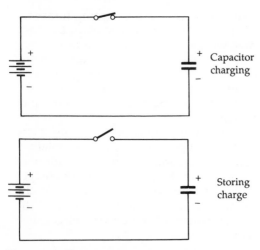

Figure 1.11 Charging a capacitor − the flow of electrons to the negative plate and from the positive plate leaves these plates charged, meaning that they each have an incorrect number of electrons. Any conducting connection between the plates will allow electrons to flow back to restore the normal balance.

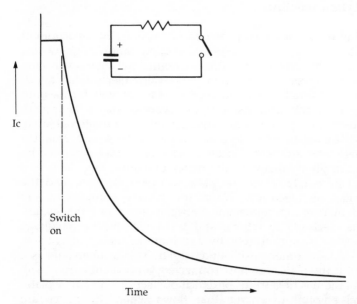

Figure 1.12 The discharge of a capacitor through a resistor follows the exponential pattern illustrated here. In theory, the discharge is never complete but for all practical purposes it is completed after a time equal to four times the time constant of R × C.

16

not constant, it follows the pattern shown in Figure 1.12. A more useful way of calculating the charge emerges from the knowledge that the quantity:

$$\frac{\text{charge on one plate}}{\text{voltage between plates}}$$

is always a constant for a particular size and arrangement of plates. This quantity is called the capacitance of the plates, and we can write the equation as:

charge = capacitance × voltage

or in symbols as:

$Q = C.V$ which can also be written as $C = Q/V$

What makes this useful is that the quantity we call capacitance can be calculated from factors such as the size and position of the plates, so that if capacitance can be calculated and voltage measured, then the amount of stored charge Q can be calculated.

When modern electrical units were being devised, the unit of charge was, as now, the coulomb, and the unit of voltage is the volt. When coulombs are divided by volts to find a unit of capacitance, this unit, called the farad, turns out to be too large for most of our practical purposes, though capacitors of 1 farad or more can be and are manufactured for specialised purposes. The usual sub-multiples that are used for the farad are the microfarad, μF, equal to 10^{-6}F; the nanofarad, nF, equal to 10^{-9}F, and the picofarad, pF, equal to 10^{-12}F.

Another point is that the larger capacitors can store a considerable amount of charge at a voltage level which can cause severe shock or burning if the terminals are touched. Even at low voltages, the amount of energy stored by a capacitor can cause destructive sparking, sufficient to melt metal wires, if a charged capacitor is short circuited. The amount of charged energy is found from:

energy = $0.5 \ CV^2$

which gives energy in joules when the units of capacitance and voltage are the farad and the volt respectively. For example, a

250µF capacitor charged to 400V will carry a stored energy of:

$250 \times 10^{-6} \times (400)^2$ which is 40 joules

and this is the amount of energy that would be achieved by a power of 40W acting for one second. When a capacitor is short circuited, however, this energy can be discharged in a fraction of a second, and any brief release of such a substantial amount of energy can be very destructive.

The electrolytic capacitor is almost a subject by itself, and it has to be treated separately from all other capacitor types because its principles are more like those of an electrolytic cell (see Chapter 2) than those of conventional capacitors. The principle is that several metals, notably aluminium and tantalum, can have very thin films of their respective oxides formed on the surface when a voltage is applied in the correct polarity (metal positive) between the metal and a slightly acidic or alkaline liquid. These very thin films then insulate the metal from the conducting liquid, the electrolyte, forming a capacitor, an electrolytic capacitor. The name comes from the resemblance to an electrolytic (metal plating) cell. Because the insulating films are very thin and of very large surface area, the capacitance value can be very large for a component that is physically small.

The most common type of electrolytic capacitor makes use of aluminium foil, which can be dimpled or corrugated to increase the effective area, enclosed in an aluminium can which is filled with a slightly alkaline solution of ammonium perborate in jelly form. The capacitor is formed by applying a slowly-rising voltage to the capacitor, with the foil positive and the case negative until the voltage reaches its rated level and the DC current falls to a minimum, indicating that the insulation is as good as it is ever likely to be. From then on, when the capacitor is used, it must have a DC (polarising) voltage applied in the same polarity so as to maintain the insulating film. If the capacitor is used with the voltage reversed, the film will be dissolved, removing any insulation and allowing large currents to pass through the liquid, which will vaporise, destroying the can. The electrolyte is usually in jelly form, but the devastation that can be caused by an exploding electrolytic (not to mention the noise) ensures that no-one who has achieved this is willing to try again.

The use of tantalum as the metal of an electrolytic allows for a very different form of construction, in which the oxide film is stable and able to withstand moderate reversals of voltage. Tantalum capacitors can be used without a steady polarising voltage, can be run with the electrolyte virtually dry, and have generally better characteristics than the traditional aluminium type of electrolytic. Experience with the use of tantalum has led to the development of 'dry' electrolytes for the aluminium type of electrolytic also.

Leakage

Because of the very fragile nature of the insulating film, which can be only a few atoms thick, electrolytic capacitors always suffer from a large amount of leakage, so that leakage current at rated voltage is quoted rather than power factor or dissipation factors. Leakage is often related to the capacitance value and working voltage, as the formula in Figure 1.13 shows, and several manufacturers make

$I = 0.01$ CV with C in μF and V in volts.
I in μA, minimum value 3 μA
Example: 500 μF 10 V will have calculated leakage
of $0.01 \times 500 \times 10 = 50 \mu$A
Example: 1 μF 63 V will have calculated leakage
of $0.01 \times 1 \times 63 = 0.63 \mu$A, but because of the 3 μA
minimum rule, the actual leakage will be 3 μA

Figure 1.13 A typical formula for leakage current for a range of electrolytic capacitors. This illustrates the general form of the relationship between leakage current and capacitance.

use of this or a similar formula to quote leakage values. Many manufacturers also quote a life expectancy in excess of 100,000 hours, at 40°C and rated voltage, for electrolytics, since there is still a considerable prejudice against their use for anything other than consumer electronics. Temperature ranges of −40°C to +85°C are often quoted, but considerable derating is needed at the higher temperatures, and there is a risk of freezing the jelly type of electrolyte at the lower temperatures. Some types of can incorporate vents in order to relieve excess gas pressure inside the electrolyte.

Use of electrolytic capacitors

Electrolytics are used predominantly as reservoir and smoothing capacitors for mains-frequency power supplies, so that their most important parameters, other than capacitance and voltage rating, concern the amount of ripple current that they can pass. For each capacitor the manufacturer will quote a maximum ripple current (typically at 100 or 120Hz), and also two parameters that concern the ability of the capacitor to pass current, ESR and impedance. The ESR is the effective series resistance, typically 50 milliohms, for low-frequency currents, and this value may set a limit to the ripple current that can pass; also to the effectiveness of the capacitor for smoothing. The other parameter is the effective impedance in milliohms, caused by self-inductance, measured at 10kHz and 20°C, which is used to measure how effectively the capacitor will bypass currents at higher frequencies. If an electrolytic capacitor is used in a decoupling circuit which is likely to handle a large frequency range, other capacitor types should be used to deal with frequencies higher than 10kHz, such as a polyester type for the range to 10MHz and a mica or ceramic for higher frequencies. The combination of low ESR and low self-inductance is essential for an electrolytic used in a switch-mode supply, see Chapter 5.

The general purpose type of electrolytic uses aluminium, often with a separate aluminium casing rated at 1000V insulation value. The physical form is a cylinder with tag, rod or screw connectors at one end. The capacitance range is generally very large for the lower-voltage units, up to 15,000μF for 16V working, but at the higher voltage ratings of 400V, values of 1μF to 220μF are more usual. The tolerance of value is large, −10% to +50%, and permitted ripple currents range from 1A to 7A depending on capacitor size. These quantities may appear to be very satisfactory, but you need to remember their significance. Supposing the capacitor is used with a 6V supply at its full rated ripple current of 5A and has an ESR of 50mΩ. This implies that the ripple voltage across the capacitor will be $5 \times 0.05V = 0.25V$, so that the ripple voltage will be 0.25V, a substantial fraction of the 6V supply, calling for further smoothing.

Smaller electrolytics are made for direct mounting on the circuit boards for decoupling or additional smoothing, and these are cylindrical and wire-terminated, either axial (a wire at each end) or radial (both wires at one end). The voltage range can be from 10 to 450V, with a working temperature range of −40°C to +85°C (derating advised at the higher temperatures), and with power factors that can be as low as 0.08 and as high as 0.2. The largest range of values is for the smaller working voltages, 10μF to 4700μF. The sub-miniature versions have working voltages that range from 6.3V to 63V, and with leakage current which is 3μA minimum, with the larger capacitances having leakage given by the formula: 0.01CV. For example, a 47μF, 40V capacitor would have leakage of: $0.01 \times 47 \times 40 = 18.8μA$.

A specialised wet-electrolyte type is made for the purposes of memory back-up in digital circuits. CMOS memory chips can retain data if a voltage, lower than the normal supply voltage, is maintained at one pin of the chip. The current demand at this pin is very low, and can therefore be supplied by a capacitor for considerable periods. This is not the method that is used for calculators, which use a battery (though solar-powered types usually incorporate a large-value capacitor in parallel with the solar cells), but for such devices as central-heating controllers which must retain their settings if the mains supply fails for a comparatively short period. Typical values for these electrolytics are 1F0 and 3F3. Discharge times range from 1 to 5 hours at 1mA and 300 to 500 hours at the more typical current requirement of 5μA, but the high leakage current must be taken into account.

Solid-electrolyte types are now available in the aluminium range of electrolytics. Unlike the traditional type of aluminium electrolytic, these need no venting, and cannot suffer from evaporation of the electrolyte. Also, unlike the traditional electrolytic, they can be run for periods with no polarising voltage, and can accept reverse voltage, though only about 30% of rated forward voltage at 85°C, considerably less at higher temperatures. Typical sizes are from 2.2μF to 100μF, with voltage ratings of 10V to 35V at 85°C. Temperature range is −55°C to +125°C, and even at the maximum working temperature of 125°C, the life expectancy is in excess of 20,000 hours. The leakage currents are fairly high, in the range 9 to 250μA, and the ripple current ratings are in the range 20 to 300mA. One important feature is that the specifications place no restrictions on the amount of charge or discharge current that flows in a DC circuit, provided that the working voltage is not exceeded.

Tantalum electrolytics
Tantalum electrolytics invariably use solid electrolytes along with tantalum metal, and have much lower leakage than the aluminium types. This makes them eminently suitable for purposes such as signal coupling, filters, timing circuits and decoupling as well as in power supplies. The usual forms of these electrolytics are as miniature beads or tubular axial types. The voltage range is 6.3 to 35V, with values of 1.0μF to 100μF. The temperature range is −55°C to +85°C. Tantalum electrolytics can be used without any DC bias, and can also accept a small reverse voltage, typically less than 1.0V. A minimum leakage current of 1μA is to be expected, and for the higher values of capacitance and working voltage the

leakage current is found from the capacitance × voltage factor, subject to the 1μA minimum. Power factors in the range of 0.02 to 0.2 can be expected.

2 Battery supplies

Batteries were, as has been indicated in Chapter 1, the original source of DC, and have always been an important form of power supply for electronic equipment. Historically, batteries have been in use for over two hundred years, and the problems that are encountered with one of the simplest and oldest types of battery are a good introduction to the reasons why so many diverse battery types exist nowadays, and to the technology that is used.

Strictly speaking, a battery is an assembly of single cells, so that the action of a cell is the subject of this chapter. Any type of cell depends on a chemical action which is usually between a solid (the cathode plate) and a liquid, the electrolyte. The use of liquids makes cells less portable, and the trend for many years has been to using jellified liquids, and also to materials that are not strong acids or alkalis. The voltage that is obtained from any cell depends on the amount of energy liberated in the chemical reaction, but only a limited number of chemical reactions can be used in this way, and for most of them, the energy that is liberated corresponds to a voltage of between 0.8 and 2.3V per cell with one notable exception, the lithium cell. This range of voltage represents a fundamental chemical action that cannot be circumvented by refining the mechanical or electrical design of the cell.

The current that can be obtained from a cell is, by contrast, determined by the area of the conducting plates and the resistance of the electrolyte material, so that there is a relationship between physical size and current capability. The limit to this is purely practical, because if the cell is being used for a portable piece of equipment, a very large cell makes the equipment less portable and therefore less useful.

Hundreds of types of cells have been invented and constructed since 1790, and most of them have been forgotten, even from school textbooks (though the Weston Standard Cell still occupies a place despite the fact that the more practical mercury button cell, bought from the local chemist at a tenth of the price, provides as useful a reference voltage). By the middle of this century, only one type of cell was commonly available, the Leclanché cell, which is the familiar type of ordinary torch cell. The introduction of semiconductor electronics, however, has revolutionised the cell and battery industry, and the requirements for specialised cells to use in situations calling for high current, long shelf life or miniature construction have resulted in the development and construction of cells from materials that would have been considered decidedly exotic in the earlier part of the century.

Primary and secondary cells

A primary cell is one in which the chemical reaction is not reversible. Once the cell is exhausted, because the electrolyte has dissolved all of the cathode material or because some other chemical (such as the depolariser, see later) is exhausted, then recharging to the original state of the cell is impossible, though for some types of primary cell, a very limited extension of life can be achieved by careful recharging. In general, attempts to recharge a primary cell will usually result in the internal liberation of gases which will eventually burst explosively through the case of the cell.

A secondary cell is one in which the chemical reaction is reversible. Without getting into too much detail about what exactly constitutes reversibility, reversible chemical reactions are not particularly common, and it is much more rarely that such a reaction can be used to construct a cell, so that there is not the large range of cells of the secondary type such as exists for primary cells. Even the nickel-cadmium secondary cell which is used so extensively nowadays in the form of rechargeable batteries is a development of an old design, the nickel-iron cell due to Edison in the latter years of the 19th century.

There is a third type of cell, the fuel cell, which despite very great research efforts for some thirty years has not become as common as was originally predicted. A fuel cell uses for its power a chemical reaction which is normally a combustion, the burning of a substance. This is dealt with briefly in Chapter 6.

Battery connections

When a set of cells is connected together, the result is a battery. The cells that form a battery could be connected in series, in

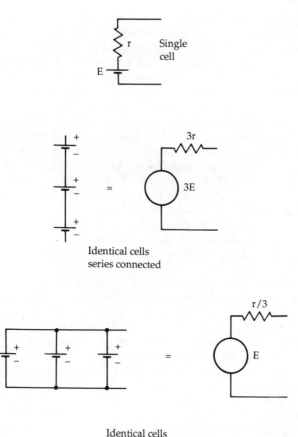

Figure 2.1 The effect of connecting cells in parallel and in series. Parallel connection is seldom used except for some secondary cells, because it can cause large currents to flow from cell to cell.

parallel, or in any of the series-parallel arrangements, but in practice the connection is nearly always in series. The effect of both series and parallel connection can be seen in Figure 2.1. When the cells are connected in series, the open-circuit voltages (EMFs) add, and so do the internal resistance values, so that the overall voltage is greater, but the current capability is the same as that of a single cell. When the cells are connected in parallel, the voltage is as for one cell, but the internal resistance is much lower, because it is the resultant of several internal resistances in parallel. This allows much larger currents to be drawn, but unless the cells each produce exactly the same EMF value, there is a risk that current will flow between cells, causing local overheating. For this reason, primary cells are never connected in parallel, and even secondary cells which are more able to deliver and to take local charging current, are seldom connected in this way except for recharging.

Higher currents are obtained by making primary cells in a variety of sizes, with the larger cells being able to provide more current, and having a longer life because of the greater quantity of essential chemicals. The limit to size is portability, because if a primary cell is not portable it has a limited range of applications. Secondary cells have much lower internal resistance values, so that if high current capability is required along with small volume, a secondary cell is always used in preference to a primary cell. One disadvantage of the usual type of nickel-cadmium secondary cell in this respect, however, is a short shelf-life, so that if equipment is likely to stand for a long time between periods of use, secondary cells may not be entirely suitable, because they will always need to be recharged just before use.

The important parameters for any type of cell are its open-circuit voltage (the EMF), its typical internal resistance value, its shelf-life, active life and energy content. The EMF and internal resistance principles have been mentioned already, and shelf life indicates how long a cell can be stored, usually at a temperature not exceeding 25°C, before the amount of internal chemical action seriously decreases the useful life. The active life is less easy to define, because it depends on the current drain, and it is usual to quote several figures of active life for various average current drain values. The energy content is defined as EMF × current × active life, and will usually be calculated from the most favourable product of current and time. The energy content is more affected by the type of chemical reaction and the weight of the active materials than by details of design.

Cell origins

All the cells that are used today can trace their origins to the voltaic pile which was invented by Alessandro Volta around 1782. Each portion of this device was a sandwich of cloth soaked in brine, Figure 2.2(a) and laid between one plate of copper and one plate of zinc. When sufficient of the sandwich cells were assembled into a battery, the voltage was enough to cause effects such as the heating of a thin wire, or the twitching of the leg of a (dead) frog

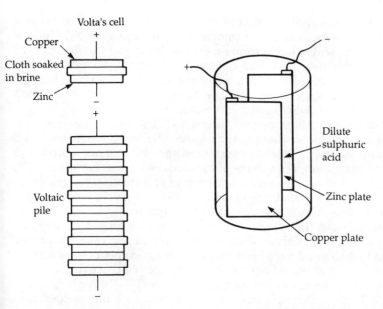

Figure 2.2 (a) Volta's original cell and the voltaic pile (battery) and (b) the original form of simple cell, which used plates of copper and of zinc dipping into dilute sulphuric acid solution.

— the effect discovered by Luigi Galvani. The sort of electrical systems we get in Italian cars came much later.

The next step was to the simple cell, as we now call it, which used the metal zinc (the cathode) and the liquid sulphuric acid to provide the chemical reaction, and the other contact, the anode, that was needed was provided by a copper plate which also dipped into the acid, Figure 2.2(b).

The action is that when the zinc dissolves in the acid, electrons are liberated. These electrons can flow along a wire connected to the zinc, and back into the chemical system through the copper plate, so meeting the requirement for a closed path for electrons. In terms of conventional current flow, a decision made long before the existence of electrons was suspected, this is a current flowing from the positive copper plate, the anode, to the negative zinc plate, the cathode. All cells conform to this pattern of a metal dissolving in an acid or alkaline solution and releasing electrons which return to the cell by way of an inert conductor which is also immersed in the solution. The original zinc/sulphuric acid type of cell is known as the simple cell to distinguish it from the many types that have followed.

The simple cell has several drawbacks that make it unsuitable for use other than as a demonstration of principles. The use of sulphuric acid in liquid form makes the cell unsuitable for any kind of portable use, since acid can spill and even at the dilution used for the simple cell it can cause considerable damage. The cell cannot be sealed, because as the zinc dissolves it liberates hydrogen gas which must be vented.

There are more serious problems. The sulphuric acid will dissolve the zinc, though at a slower rate, even when no circuit exists, so that the cell has a very short shelf life and not much active life. In addition, the voltage of the cell, which starts at about 1.5V, rapidly decreases to zero when even only a small current is taken, because the internal resistance rises to a large value as the cell is used. This makes the cell unusable until the zinc is removed, washed, and then re-inserted. No doubt if Alessandro Volta had been working with a Government grant rather than on private funds, he would have been told that his experiments were leading nowhere and would no longer be funded (and what use, pray, could ever be found for such an invention?).

The efforts that were made to understand the faults of the simple cell have led to the development of considerably better cells, because by understanding principles we are better able to design new products. The problem of the zinc dissolving even with no circuit connected was solved by using very pure zinc or by coating the zinc with mercury. The problem is one of local action, meaning that the impurities in the zinc act like anodes, forming small cells that are already short circuited. By using very pure zinc, this local action is very greatly reduced, but in the 18th century purification of metals had not reached the state that we can expect nowadays. Mercury acts to block off the impurities without itself acting as an anode, and this was a much easier method to use at the time.

The rapid increase in internal resistance proved to be a more difficult problem, and one that could not be solved other than by redesigning the cell. Curiously enough, however, the problem of the increasing internal resistance was later be used as a way of constructing electrolytic capacitors, see Chapter 1. The problem is that dissolving zinc in sulphuric acid releases hydrogen gas, and this gas coats the surface of the anode as it is formed, an action that was originally called polarisation. The gas appears at the anode because of the action of the electrons entering the solution from the external circuit. Because hydrogen is an insulator, the area of the anode that can be in electrical contact with the sulphuric acid is greatly reduced by this action, so that the internal resistance increases. When local action is present, the internal resistance will increase from the moment that the cell is assembled, though for the pure-zinc cell or the type in which the zinc has been coated (amalgamated) with mercury, the internal resistance increases only while the cell is used.

The problem can be solved only by removing the hydrogen as it forms or by using a chemical reaction that does not generate any gas, and these are the solutions that have been adopted by every successful cell type developed since the days of Volta. The removal of hydrogen is achieved by using an oxidising material, the depolariser, which has to be packed around the anode. The depolariser must be some material which will not have any chemical side-effects, and insoluble materials like manganese (II) oxide have been used very successfully in the past and are still widely used.

The Leclanché cell

The cell that was developed by the French chemist Leclanché in the 19th century has had a remarkably long history, and in its dry form is still in use, though now grandified by the title of carbon-zinc cell. In its original form, Figure 2.3, the electrolyte was a liquid, a solution of ammonium chloride. This is mildly acid, but not corrosive in the way that sulphuric acid is, and one consequence of using this less acidic electrolyte is that the zinc, even if not particularly pure, does not dissolve in the solution to the same extent when no current is passing in the external circuit. Local action is still present, but greatly reduced as compared to a zinc-acid type of cell.

The anode for the cell is a rod of carbon, a material which is chemically inert and therefore unattacked by the electrolyte. The carbon rod is surrounded by a paste of manganese dioxide, all contained inside a porous pot so that the electrolyte keeps the whole lot wet and conducting. The action when current flows is that zinc dissolves in the mildly acid solution, releasing electrons which then travel through the circuit. At the anode, the electrons would normally react with the water in the liquid to produce hydrogen, but the action of the manganese dioxide is to absorb electrons in preference to allowing the reaction with the water to proceed, producing a different oxide of manganese (a reduced state). As the cell operates, the zinc is consumed, as also is the manganese dioxide, and when either is exhausted the cell fails. The open-circuit voltage is about 1.5V, and the internal resistance can be less than one ohm.

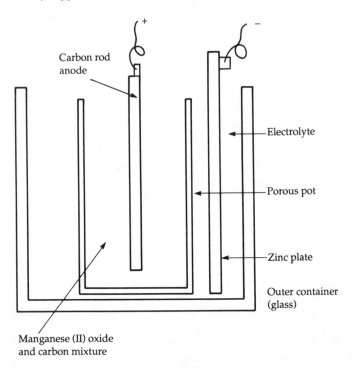

Figure 2.3 The original form of the Leclanché wet cell, using ammonium chloride as the electrolyte and with a manganese (II) oxide depolariser. The porous pot is used to separate the chemicals while allowing ions to move in the solution.

The older form of the Leclanché cell was in service for operating doorbells and room indicators from mid-Victorian times, and some that had been installed in these days were still working in the late 1930s. The reason is that the Leclanché cell was quite remarkably renewable. The users could buy spare zinc plates, spare ammonium chloride (which could also be used for smelling-salts) and spare manganese dioxide, so that the cell could be given an almost indefinite life on the type of intermittent use that it had. Some worked for well over 20 years without any attention at all, tucked away in a cupboard on a high shelf.

The dry form of the Leclanché cell is the type that until quite recently was the only familiar form of primary cell. The construction, Figure 2.4, follows the principles of the older wet type of cell, but the ammonium chloride electrolyte is in jelly form rather than liquid, and the manganese oxide is mixed with graphite and with

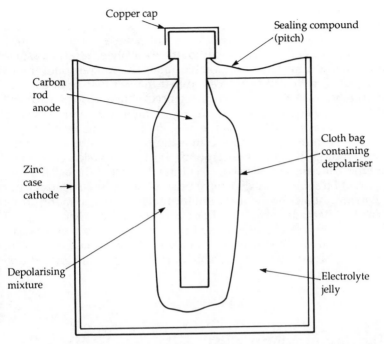

Figure 2.4 The familiar dry form of the Leclanché cell, now known as the carbon-zinc cell.

some of the jelly to keep it also moist and conducting. The action is the same, but because the dry cell is usually smaller than the wet variety and because its jelly electrolyte is less conductive, this form of the cell has generally a higher internal resistance than the old wet variety. The advantage of portability, however, totally overrules any disadvantages of higher internal resistance, making this the standard dry cell for most of the twentieth century.

The carbon-zinc dry cell, as it is more often called now, fails either when the zinc is perforated or when the manganese dioxide is exhausted. One of the weaknesses of the original design is that the zinc forms the casing for the cell, so that when the zinc becomes perforated, the electrolyte can leak out, and countless users of dry cells will have had the experience of opening a torch or a transistor radio battery compartment to find the usual sticky mess left by leaking cells. The term dry cell never seems quite appropriate in these circumstances.

The problem can not simply be dealt with by using a thicker zinc casing and by restricting the amount of manganese dioxide so that the cell fails because of high internal resistance before the zinc is used up. The carbon-zinc cell does not have a particularly long shelf life and once it has been used, the electrolyte starts to dissolve the zinc at a slow but inexorable rate. This corresponds to an internal current within the cell, called the self-discharge current. Perforation will therefore invariably occur when an exhausted cell is left inside equipment, and the higher the temperature at which the cell is kept, the faster is the rate of attack of the zinc.

This led to the development of leakproof cells with a steel liner surrounding the zinc. Leakproofing in this way allowed a much thinner zinc shell to be used, so cutting the cost of the cell (though it could be sold at a higher price because of the leakproofing) and allowing the cell to be used until a much greater amount of the zinc had been dissolved. Leakproofing is not foolproof, and even the steel shell can be perforated in the course of time, or the seals can fail and allow electrolyte to spill out. Nevertheless, the use of the steel liner has considerably improved the life of battery-operated equipment.

The alkaline primary cells

A different group of cell types makes use of alkaline rather than acid electrolytes, so that though the principle of a metal dissolving in a solution and releasing electrons still holds good, the detailed chemistry of the reaction is quite different. On the assumption that the reader of this book will be considerably more interested in the electrical characteristics of these cells rather than the chemistry, we will ignore the chemical reactions unless there is something about them that requires special notice. One point that does merit attention is that the alkaline reactions do not generate gas, and this allows the cells to be much more thoroughly sealed than the zinc-carbon type. It also eliminates the type of problems that require the need of a depolariser, so that the structure of alkaline cells can, in theory at least, be simpler than that of the older type of cell. Any attempt to recharge these cells, however, will generate gas and the pressure will build up until the container fractures explosively.

The best-known alkaline type of cell is the manganese alkaline, whose construction is illustrated in Figure 2.5. This was invented by Sam Ruben in 1939 and was used experimentally in some

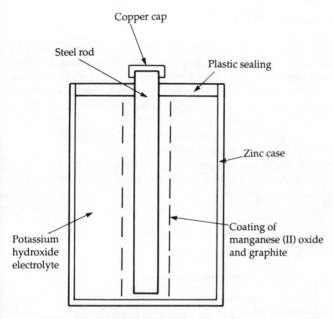

Figure 2.5 The construction and composition of a manganese alkaline cell, which has a much longer shelf life and energy content than the carbon-zinc type.

wartime equipment, but the full-scale production of manganese-alkaline cells did not start until the 1960s. The cell uses zinc as the cathode, with an electrolyte of potassium hydroxide solution, either as liquid or as jelly, and the anode is a coating of manganese (II) oxide mixed with graphite and laid on steel. The cell is sealed because the reaction does not liberate gas, and the manganese (II) oxide is being used for its manganese content rather than for its oxygen content as a depolariser. The EMF of a fresh cell is 1.5V, and the initial EMF is maintained almost unchanged for practically the whole of the life of the cell. The energy content, weight for weight, is higher than that of the carbon-zinc cell by a factor of 5–10, and the shelf life is very much better due to an almost complete lack of secondary action. All of this makes these cells very suitable for electronics use, particularly for equipment that has fairly long inactive periods followed by large current demand. Incidentally, though the cells use alkali rather than acid, potassium hydroxide is a caustic material which will dissolve the skin and is extremely dangerous to the eyes. An alkaline cell must never be opened, nor should any attempt ever be made to recharge it.

Miniature (button) cells

The miniature cells are the types specified for deaf-aids, calculators, cameras and watches, but they are quite often found in other applications, such as for backup of memory in computing applications and for smart-card units in which a credit-card is equipped with a complete microprocessor and memory structure so that it keeps track of transactions. The main miniature cells are silver-oxide and mercury, but the term mercury cell can be misleading, because metallic mercury is not involved.

The mercuric oxide button cell, to give it the correct title, uses an electrolyte of potassium hydroxide (Figure 2.6) which has had

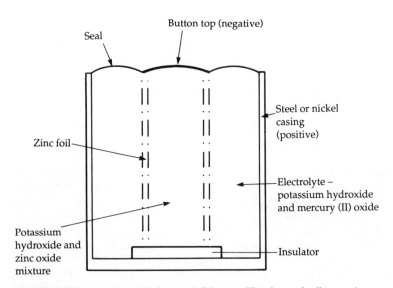

Figure 2.6 The mercuric oxide button cell layout. This form of cell is used extensively in miniature form.

zinc oxide dissolved in it until saturated, so that the cell can be classed as an alkaline type. The cathode is the familiar zinc, using either a cylinder of perforated zinc foil or a sintered zinc-powder cylinder fastened to the button-top of the cell and insulated from the bottom casing. The anode is a coating of mercury (I) oxide mixed with graphite to improve conductivity and coated on nickel-plated steel or stainless steel which forms the casing of the cell.

The EMF of such cells is low, 1.2–1.3V, and the energy content is high, with long shelf-life due to the absence of local action.

The silver-oxide cell is constructed in very much the same way as the mercuric oxide cell, but using silver (I) oxide mixed with graphite as the anode. The cathode is zinc and the electrolyte is potassium hydroxide as for the mercuric oxide cell. The EMF is 1.5V, a value which is maintained at a steady level for most of the long life of the cell. The energy content is high and the shelf life long.

All of these miniature cells are intended for very low current applications, so that great care should be taken to avoid accidental discharge paths. If the cells are touched by hand, this will leave a film of perspiration which is sufficiently conductive to shorten the life of the cell drastically. When these cells are fitted, they should be moved and fitted with tweezers, preferably plastic tweezers, or with dry rubber gloves if you need to use your hands. These cells should not be recharged, nor disposed of in a fire. The mercury type is particularly hazardous if mercury compounds are released, and they should be returned to the manufacturer for correct disposal if this is possible, otherwise disposed of by a firm that is competent to handle mercury compounds.

Lithium cells

Lithium is a metal akin to potassium and sodium, and it is highly reactive, so much so that it cannot be exposed to air, and it reacts with explosive violence when put into contact with water. The reactive nature of lithium metal means that a water solution cannot be used as the electrolyte, and much research has gone into finding liquids which ionise to some extent but which do not react excessively with lithium. A sulphur-chlorine compound, thionyl chloride, is used, with enough dissolved lithium salts to make the amount of ionisation sufficient for the conductivity that is needed. The lithium, (Figure 2.7) is coated on to a stainless-steel mesh which is separated from the rest of the cell by a porous polypropylene container. The anode is a mixture of manganese (II) oxide and graphite, also coated on to stainless-steel mesh. The whole cell is very carefully sealed.

The reaction can be used to provide a cell with an exceptionally high EMF of 3.7V, very long shelf life of 10 years or more, and high energy content. The EMF is almost constant over the life of the cell, and the internal resistance can be low. Lithium cells are

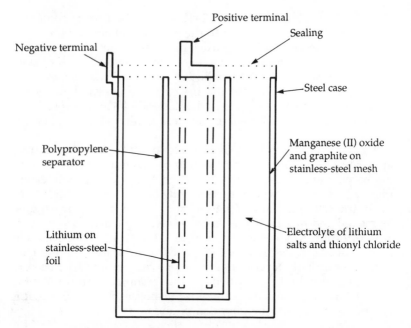

Figure 2.7 The lithium cell, used in electronics for backup of memory banks, since it has a high output voltage and a very long shelf life, with a high energy content. The presence of lithium makes it very dangerous to pierce this cell.

expensive, but their unique characteristics have led to their being used in automatic cameras, where focusing, film wind, shutter action, exposure and flash are all dependent on one battery, usually a two-cell lithium type.

For electronics applications, lithium cells are used mainly for memory backup, and very often the life of the battery is as great as the expected lifetime of the memory itself. The cells are sealed, but since excessive current drain can cause a build-up of hydrogen gas, a safety-valve is incorporated in the form of a thin section of container wall which will blow out in the event of excess pressure. Since this will allow the atmosphere to reach the lithium, with risk of fire, the cells should be protected from accidental over-current which would cause blow-out. A recommended protection circuit is illustrated in Figure 2.8. This is for use in applications where the lithium cell is used as a backup, so that D1 conducts during normal memory operation and D2 conducts during backup. Short-circuit failure of D2 would cause the lithium cell to be charged by the

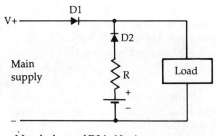

Max. leakage of D2 is 10 μA
R protects against breakdown of D2

Figure 2.8 A recommended protection circuit for a lithium cell in a simple backup action. This prevents reverse current from damaging the cell.

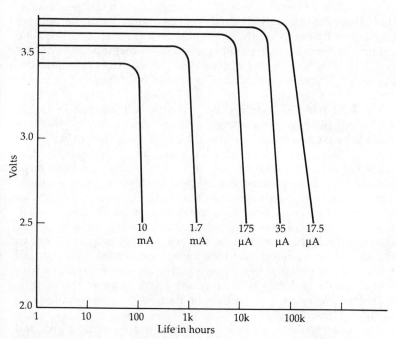

Figure 2.9 A typical graph of battery voltage, current and life for a lithium battery at normal room temperatures.

37

normal supply, and the resistor R will then limit the current to an amount which the cell manufacturer deems to be safe. If the use of a resistor would cause too great a voltage drop in normal backup use, it could be replaced by a quick-blowing fuse, but this has the disadvantage that it would cause loss of memory when the main supply was switched off.

Lithium cells must never be connected in parallel, and even series connection is discouraged and limited to a maximum of two cells. The cells are designed for low load currents, and Figure 2.9 shows a typical plot of battery voltage, current and life at 20°C. Some varieties of lithium cells exhibit voltage lag, so that the full output voltage is available only after the cell has been on load for a short time — the effect becomes more noticeable as the cell ages. Another oddity is that the capacity of a lithium cell is slightly lower if the cell is not mounted with the positive terminal uppermost.

Secondary cells

A secondary cell makes use of a reversible chemical process, so that when the cell is discharged, reverse current into the cell will recharge it by restoring the original chemical constitution. Unlike primary cell reactions, reversible reactions of this type are unusual and only two basic types are known, the lead-acid type and the alkali-metal type, both of which have been used for a considerable time.

The lead-acid cell construction principle is illustrated in Figure 2.10. Both plates are made from lead and are perforated to allow them to be packed with the active materials. One, the positive plate (anode), is packed with lead (IV) oxide, and the negative plate (cathode) is packed with spongy or sintered lead which has a large surface area. Both plates are immersed in sulphuric acid solution. The acidity is much greater than that of the electrolytes of any of the acidic dry cells, and very great care must be taken when working with lead-acid cells to avoid any spillage of acid or any charging fault that could cause the acid to boil or to burst out of the casing. In addition, the recharging of a vented lead-acid cell releases hydrogen and oxygen as a highly explosive mixture which will detonate violently if there is any spark nearby. The EMF is 2.2V (nominally 2.0V), and the variation in voltage is quite large as the cell discharges.

The older vented type of lead-acid cell is now a rare sight, and modern lead-acid cells are sealed, relying on better control of charging equipment to avoid excessive gas pressure. The dry type

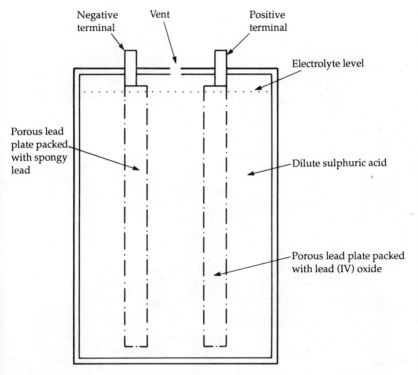

Figure 2.10 The principles of construction of a lead-acid secondary cell. Modern cells can be made lighter and more compact than the older types by using interleaving plates of sintered construction, and better control of charging has led to improved reliability.

of cell uses electrolyte in jelly form so that these cells can be used in any operating position. Cells that use a liquid electrolyte are constructed with porous separator material between the plates so that the electrolyte is absorbed in the separator material, and this allows these cells also to be placed in any operating position. Since gas pressure build-up is still possible if charging circuits fail, cells are equipped with a pressure operated vent which will reseal when pressure drops again.

Lead-acid cells are used in electronics applications mainly as backup power supplies, as part of uninterruptable power systems, where their large capacities and low internal resistance can be utilised. Capacity is measured in ampere-hours (Ah), and sizes of

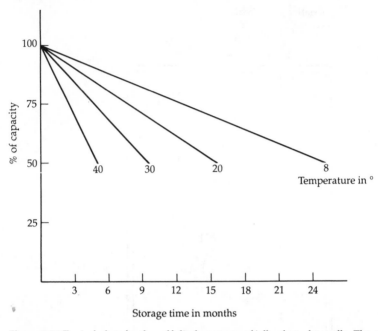

Figure 2.11 Typical plots for the self-discharge rate of jelly-electrolyte cells. The rate can depend considerably on the construction methods used for the cell.

9Ah to 110Ah are commonly used. Care should be taken in selecting suitable types — some types of lead-acid cells will self-discharge considerably faster than others and are better suited to applications where there is a fairly regular charge/discharge cycle than for backup systems in which the battery may be used only on exceptional occasions and charging is also infrequent. Figure 2.11 shows the self discharge rates of jelly-electrolyte cells at various temperatures, taking the arbitrary figure of 50% capacity as the discharge point.

Lead-acid batteries need to be charged from a constant-voltage source of about 2.3V per cell at 20°C— Figure 2.12 shows the variation of charging voltage per cell with ambient temperature of the cell. Cells can be connected in series for charging provided that all of the cells are of the same type and equally discharged. A suitable multi-cell charger circuit is illustrated in Figure 2.13. For batteries of more than 24V (12 cells) the charging should be in 24V blocks, or a charging system used that will distribute charging so

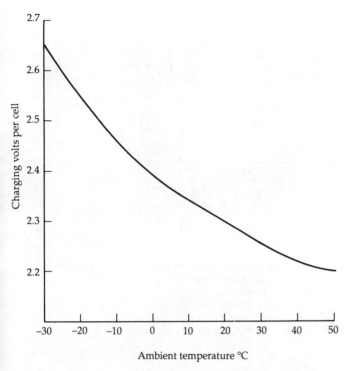

Figure 2.12 The way that charging voltage per cell for a lead-acid battery will vary with temperature.

that no single cell is being over-charged. Parallel charging may be used if the charger can provide enough current.

The operating life of a lead-acid cell is usually measured in terms of the number of charge/discharge cycles, and is greater when the cell is used with fairly high discharge currents — the worst operating conditions are of slow discharge and erratic re-charge intervals, the conditions that usually prevail when these cells are used for backup purposes. One condition to avoid is *deep discharge*, when the cell has been left either on load or discharged for a long period. In this state, the terminal voltage falls to 1.6V or less and the cell is likely to be permanently damaged unless it is immediately re-charged at a very low current over a long period. Typical life expectancy for a correctly operated cell is of the order of 750–6000 charge/discharge cycles.

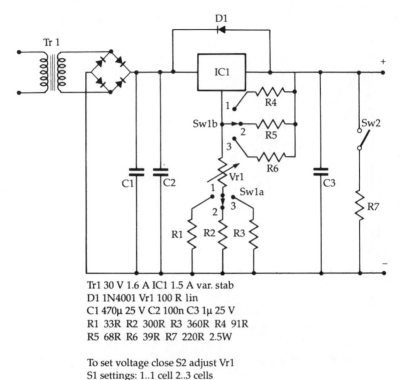

Tr1 30 V 1.6 A IC1 1.5 A var. stab
D1 1N4001 Vr1 100 R lin
C1 470μ 25 V C2 100n C3 1μ 25 V
R1 33R R2 300R R3 360R R4 91R
R5 68R R6 39R R7 220R 2.5W

To set voltage close S2 adjust Vr1
S1 settings: 1..1 cell 2..3 cells
            3..6 cells

Figure 2.13 A multi-cell charger circuit (courtesy of RS Components Ltd.)

Nickel-cadmium cells

The original type of alkaline secondary cell, invented by Edison at
the turn of the century, was the nickel-cathode iron-anode type,
using sodium hydroxide as the electrolyte. The EMF is only 1.2V,
but the cell can be left discharged for long periods without harm,
and will withstand much heavier charge and discharge cycles than
the lead-acid type. Though the nickel-iron alkaline secondary cell
still exists powering milk-floats and fork-lift trucks, it is not used
in the smaller sizes because of the superior performance of the
nickel-cadmium type of cell which is now the most common type
of secondary cell used for cordless appliances and in electronics
uses.

Nickel-cadmium cells can be obtained in two main forms, mass-plate and sintered plate. The mass-plate type uses nickel and cadmium plates made from smooth sheet, the sintered type has plates formed by moulding powdered metal at high temperatures and pressures, making the plates very porous and of much greater surface area. This makes the internal resistance of sintered-plate cells much lower, so that larger discharge currents can be achieved. The mass-plate type, however, has a much lower self-discharge rate and is more suitable for applications in which re-charging is not frequent. Typical life expectancy is from 700 to 1000 charge/discharge cycles.

One very considerable advantage of the nickel-cadmium cell is that it can be stored for 5 years or more without deterioration. Though charge will be lost, there is nothing corresponding to the deep discharge state of lead-acid cells which would cause irreversible damage. The only problem that can lead to cell destruction is reverse polarity charging. The cells can be used and charged in any position, and are usually supplied virtually discharged so that they must be fully charged before use. Most nickel-cadmium cell types have a fairly high self-discharge rate, and a cell will on occasions refuse to accept charge until it has been reformed with a brief pulse of high current. Cells are usually sealed but provided with a safety-vent in case of incorrect charging.

In use, the nickel-cadmium cell has a maximum EMF of about 1.4V, 1.2V nominal, and this EMF of 1.2V is sustained for most of the discharge time. The time for discharge is usually taken arbitrarily as the time to reach an EMF of 1V per cell, and Figure 2.14 shows typical voltage-time plots for a variety of discharge rates. These rates are noted in terms of capacity, ranging from one tenth of capacity to unity capacity, when capacity is in ampere-hours and discharge current in amperes. For example, if the capacity is 10Ah, then a C/5 discharge rate means that the discharge current is 2A.

Charging of nickel-cadmium cells must be done from a constant-current source, in contrast to the constant-voltage charging of lead-acid types. The normal rate of charge is about one tenth of the ampere-hour rate, so that for a 20Ah cell, the charge rate would be 2A. Sintered types can be recharged at higher rates than the mass-plate type, but the mass-plate type can be kept on continuous trickle charge of about 0.01 of capacity (for example, 10mA for a cell of 1Ah capacity). At this rate, the cells can be maintained on charge for an extended period after they are fully charged, but this over-charge period is about three times the normal charging time. Equipment such as portable and cordless phones which would

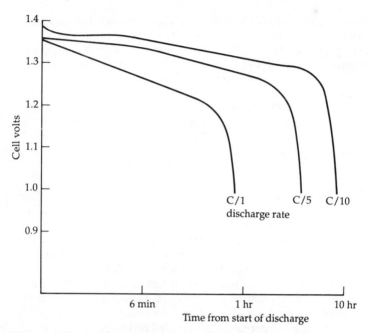

Figure 2.14 Typical characteristics for small nickel-cadmium cells.

otherwise be left on charge over extended intervals such as holiday weekends and office holidays should be disconnected from the charger rather than left to trickle-charge. This means that a full charge will usually be needed when work resumes, but the life of the cells can be considerably extended if the very long idle periods of charging can be avoided. Another option is to leave the equipment switched on so as to discharge the cells, and fit the mains supply with a timer so that there will periodic recharging.

Figure 2.15 shows a recommended circuit for recharging. This uses a 7805 regulator to provide a fixed voltage of 5V across a resistor, so that the value of the current depends on the choice of resistor and not on the voltage of the cell. The value of the resistor has to be chosen to suit the type of cell being recharged; values from 10R to 470R are used depending on the capacity of the cell. Because the regulator system is floating with respect to earth, this can be used for charging single cells or series sets of a few cells. Ready-made chargers are also available which will take various cells singly or in combination, with the correct current regulation for each type of cell.

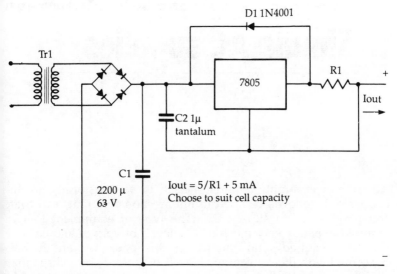

Figure 2.15 A recommended charging circuit for nickel-cadmium cells (courtesy of RS Components.)

A form of silver cell has also been used in rechargeable form. This uses an anode of porous zinc, usually a sintered component, with a silver (I) oxide and graphite cathode. The electrolyte is potassium hydroxide solution which has been saturated with zinc hydroxide. The cell can take a limited number of recharging cycles, but is now uncommon.

3 Simple AC supplies

An AC power supply requires the AC to be transformed to the correct voltage level, rectified to unidirectional current, and then smoothed so as to supply DC. The type of equipment that is required depends very much on the levels of voltage and current that are required, and though in this book we are mainly concerned with the conventional levels of current and voltage that are involved in small-scale electronics equipment, a glance outside this limited range is useful now and again.

At low levels, meaning voltages of up to 50V and currents of up to a few amperes, the standard methods make use of silicon junction diodes, usually in bridge form, and electrolytic capacitors, with AC being provided by way of small transformers. Higher-current supplies demand diodes mounted on heat sinks, along with the use of Schottky diodes which have lower forward voltage drops, and higher voltage levels are catered for by silicon diodes up to considerable current levels. Specialised EHT silicon diodes can be used for voltages as high as 7kV RMS per diode. The use of semiconductor diodes becomes prohibitive only when the voltage and the current are both very high, in the kilovolt and kiloampere range, when the older technology of mercury-vapour rectifiers is still used. For some applications, particularly where temperatures are high, vacuum valve diodes are still in use. At the higher voltage levels, the use of electrolytic capacitors for smoothing must be ruled out, and plastic rolled capacitors or oil-filled paper types have to be used, with a considerable penalty in terms of bulk.

Rectifiers

The most common form of rectifier device is the silicon diode, whose symbol and characteristics are illustrated in Figure 3.1. The

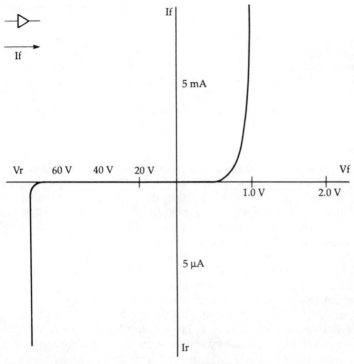

Figure 3.1 Silicon diode symbol and typical characteristics. The reverse characteristic has to be plotted with a different scale of voltage and current because reverse resistance is very high.

threshold voltage, below which no current flows in the forward direction, is about 0.6V and is a quantity that is determined by the nature of silicon and the use of a PN junction rather than by construction. Once conducting, the shape of the V/I characteristic is exponential so that a considerable increase in current is required to increase the voltage drop across the diode, as if the diode had a variable internal resistance whose value decreased as the current increased. Manufacturers will usually specify the value of forward voltage at various values of forward current, or quote an 'average' value of slope resistance.

In the reverse direction, the silicon diode is non-conductive with negligible leakage current until the reverse breakdown voltage is reached. At the reverse breakdown voltage, the amount of current that can flow is controlled only by the total resistance in the circuit, so that reverse breakdown is usually destructive, causing enough current to flow to overheat the diode and burn it out. Because of the importance of avoiding reverse breakdown, diodes are usually rated for V_{RRM}, the maximum reverse voltage that will be applied. This is not the same as the RMS voltage of the AC that is being rectified, and its value depends on the type of rectifying circuit that is used, see later in this chapter.

Because of the forward voltage, a silicon diode dissipates power, and the amount of power in watts is given by:

$$P = V_f \times I_f$$

where V_f is in volts (usually around 0.8V) and I_f is in amperes. For example, if the forward voltage at 5A is 0.8V, then the power dissipated is $0.8 \times 5 = 4.0W$. The diode must be capable of dissipating this amount of power, otherwise overheating will occur and a heatsink will have to be used, requiring the silicon diode to be of the stud-mounted variety. Diode bridge assemblies can be mounted on to heatsinks of the patterns shown in Figure 3.2. In addition, care must be taken that the peak ratings are not exceeded. A diode may be used at a current which on average is within the rated value but which consists of current peaks which might be on the limit of the ability of the diode to withstand. Most silicon diodes can withstand very large current peaks provided that these are brief, but where very low transformer winding resistance values exist and the diode is connected directly to a low-resistance electrolytic (see later for capacitor ESR), then it is possible to exceed the peak current rating of a diode, causing breakdown and failure.

The normal junction silicon diode is often replaced for high-current rectification applications by Schottky diodes. The conventional silicon diode uses a rectifying junction between P-type and N-type silicon, but the Schottky diode uses N-type silicon in a junction with metallic aluminium. Unlike the PN junction, a Schottky junction is homopolar, meaning that only one polarity of carrier (electrons in this case) is used. For mains-frequency rectification, a more important effect is that the forward voltage is significantly lower even at high currents. Typical examples from the Motorola range include diodes with forward voltage levels of 0.475V at 3A, up to 0.78V at 300A for a matched pair of high-current diodes. These should be compared with voltage levels of 1.4–1.6V for silicon junction diodes of the same current capabilities.

High-vacuum (valve) diodes are used mainly for very high-voltage applications, many of which also involve large currents, such as supplies for X-ray equipment which might require 200kV

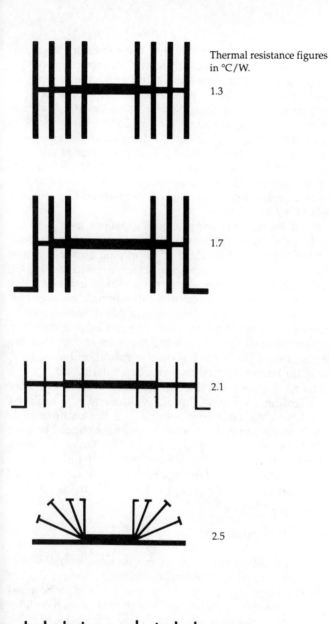

Thermal resistance figures in °C/W.

1.3

1.7

2.1

2.5

3.0

Figure 3.2 Typical heatsink shapes that can be used for stud-mounted diodes, or diode bridge assemblies.

49

at 1A. At these voltage and current levels, the valves are large and the insulation requirements are stringent. All vacuum diodes involve the emission of electrons from a hot wire, the cathode or filament, and for high-voltage rectifiers the filaments will be tungsten or (particularly for X-ray supplies) platinum. A typical filament supply would require some 110V at 300A. There is only a negligible forward voltage (compared to the output voltage level) required for conduction, but at the levels of current that are used for these rectifiers there may be anything from 50V to 150V across the valve itself during forward conduction. The outstanding feature of the vacuum diode is its ability to withstand high reverse voltage levels of 500kV or more.

Some types of equipment require high currents in the 1000A level upwards at voltage levels of 25kV or more. Mercury-arc rectifiers are the traditional method of supplying this need, and are still to be found, although semiconductors are encroaching on these applications. The mercury-arc rectifier makes use of a valve construction in which the presence of liquid mercury ensures that a vacuum is not present and the valve is filled with mercury vapour. This ionises and the ions carry current in one direction only. A filament is used for pre-heating the mercury so that the vapour is at the correct pressure, but once rectification has started, the filament is no longer needed and can be switched off.

Old equipment, particularly older radio and TV receivers, may still contain selenium rectifiers. These should always be replaced by silicon diodes in the course of servicing, if the equipment is likely to remain in use, and the selenium rectifiers should be sent to a specialist for recovery of the selenium. These rectifiers provided a long and useful life, but when they failed and selenium was burned, the fumes were dangerous. For safety reasons they should be replaced on any equipment that is likely to be used.

Silicon diode configurations

For single-phase circuits, silicon diodes can be used in half-wave, bi-phase half-wave and full-wave bridge circuits, Figure 3.3, of which the full-wave bridge is the circuit most likely to be used other than for low-current applications. The output of each rectifier circuit is a set of half-waves of roughly sine shape and since this supply is unidirectional it can be used as it is for some purposes. This makes it useful to know the value of the average output voltage in DC terms.

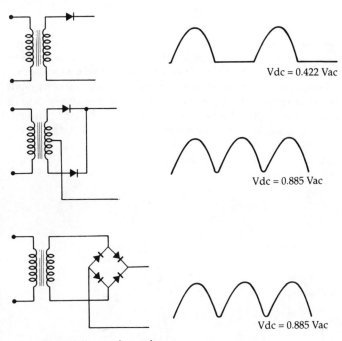

Vdc = 0.422 Vac

Vdc = 0.885 Vac

Vdc = 0.885 Vac

Note: Vac is full secondary voltage
ignoring any centre-tap. Diode drops
are also ignored

Figure 3.3 The three main silicon diode rectifier circuits for single-phase rectification.

For a half-wave rectifier, if we neglect the voltage drop across the diode, the average level of output is $0.442 \times V_{ac}$, where V_{ac} is the RMS value of the AC output from the transformer. For a 10V RMS AC output, for example, the equivalent DC is 4.42V. Since this is unsmoothed it is unsuitable for most electronics applications but it can be used as a supply for non-critical motors or even for some battery-charging applications. The important point about this value is that it is the equivalent amount of DC that is still present when all smoothing fails, and is the value to which the voltage will fall when excessive current is taken from a half-wave smoothed supply. When the diode drop is considered, it makes the level smaller by about this amount, and if a further voltage drop is caused by the resistance of transformer windings, connections, fuses etc. then these drops also have to be subtracted from the output level. The DC average output current is about $0.637 \times I_{ac}$ where I_{ac} is the RMS value of current supplied by the transformer.

The bi-phase half-wave type of rectifier requires a centre-tapped transformer winding, or a transformer with two identical secondary windings that can be connected in phase to act as a tapped winding. Care should be taken not to connect such windings out of phase so that their voltages oppose each other — always check the open-circuit output voltage with an AC meter before connecting such a transformer to the rectifiers. The current paths in the

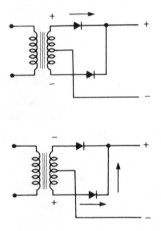

Figure 3.4 Figure The current paths in the bi-phase half-wave circuit for the two conducting phases.

two phases of conduction are illustrated in Figure 3.4, with the diodes conducting alternately. With this type of rectification, once again neglecting diode and resistive voltage drops, the output voltage DC equivalent is 0.885 of the RMS AC voltage, so that a 10V RMS transformer output would give an 8.85V equivalent DC output. The improvement is due to the bi-phase rectification, and once again, this can often be used as it is for motors or for some forms of battery charging. The average value of DC output current is 0.885 of the RMS AC current value.

When the diode drops are taken into account, the reduction in output level is only one diode drop (usually 0.6 — 0.8V) lower, and only one set of transformer winding resistance (one half of the tapped winding) need be taken into consideration when calculating resistive drops. Any inaccuracy in the tapping position of the transformer will cause one half-cycle to be of greater amplitude than the other half-cycle, and this will introduce an additional hum component (at mains frequency) into the output.

The predominant rectifier circuit for electronics use is the full-wave bridge type. This is reflected in components catalogues which list ready-made diode bridges in a wide range of voltage and current levels. Typically these range from 50V peak reverse to 1200V, and from under 1A to 60A, covering the vast majority of electronics requirements. A single transformer winding is used, but the output current always flows through two diodes, so that the voltage drop due to the diodes is twice as great as for the other two circuits. The current paths are shown in Figure 3.5, with the

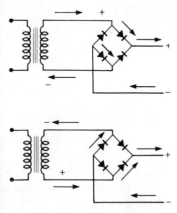

Figure 3.5 The current paths in a bridge rectifier circuit, showing that the current path is always through two diodes, making the diode drop of the order of 1.5V at least.

circuit drawn in the conventional diamond shape that is also used as a circuit symbol. The DC output level average voltage is 0.885 of AC RMS output, and average DC current is also 0.885 of RMS AC current in the winding, neglecting diode drops and drops across circuit resistances. The typical level of diode drop is around 1.5V because of the current path through two diodes in each phase of conduction.

Both single-rectifier and full-wave bridge circuits can be used with three-phase supplies, and in each case the efficiency of rectification is considerably improved. Figure 3.6 shows one form of half-wave three-phase circuit for which the average DC level is $1.17 \times V_{ac}$ (per phase) with average DC current equal to 1.73 of RMS current in each phase. The bridge equivalent was shown in

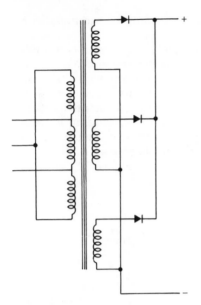

Figure 3.6 A Three-phase half-wave rectifier circuit which gives much lower ripple than a single-phase half-wave circuit.

Figure 1.10, and this gives an average DC voltage of $1.35 \times V_{ac}$ per phase, and current of 1.22 times AC RMS in each phase. The output from the bridge circuit is notably smooth, with ripple content of only 0.042 of peak DC output as opposed to the 0.48 of output for a single-phase bridge circuit when each is loaded by an inductor. Three-phase power supplies, however, are not common for electronics use.

Waveform and smoothing

The waveform from a full-wave bridge or bi-phase half-wave rectifier circuit is, if diode drops are neglected, of the form shown in Figure 3.3, a set of half-cycles in one direction. This already has a considerable DC content, but also has an unacceptable high ripple (alternating content) for most electronics applications. For some thyristor circuits, however, a raw rectified supply is essential in order to allow the thyristor to attain a non-conducting state.

The most common method of smoothing the output from such a rectifier is by the use of a large reservoir capacitor. The name aptly describes the action. The reservoir capacitor is charged as the diodes conduct, reaching the peak DC value (less diode voltage drops). As the output waveform drops to zero again, the diodes will cut off, leaving the reservoir capacitor storing the peak voltage value. If there is no leakage current, the capacitor will remain storing this voltage, and the diodes will not conduct again — in practice, with no load, the diodes will conduct sufficiently to keep the capacitor charged to the peak voltage of the AC output from the transformer, Figure 3.7.

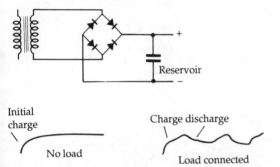

Figure 3.7 The reservoir capacitor will charge in the first conducting half-cycle from the rectifiers to a level equal to peak voltage (less diode drop).

This has two effects. One is that the diodes conduct briefly at around the peak voltage, so that the diode current is flowing for very much less than the half-cycle that is used in the un-smoothed circuit. The other effect is that the reverse voltage across each diode is much greater. In a bi-phase half-wave circuit, the voltage across the capacitor is equal to the AC peak, and the diode can also have the AC negative peak on its anode, so that the total voltage across the diode is equal to twice the peak voltage of the AC. In the full-wave bridge circuit, this reverse voltage affects two diodes in series, and if the reverse leakage currents are matched, each diode will experience only the AC peak voltage. It would, however, be rather foolhardy to assume that the reverse voltage always affects the two diodes exactly equally.

When a load is connected, the situation becomes considerably more complex. Unless the load current is almost negligible, the addition of a load will cause the output DC voltage level to drop, and the diodes will conduct between this level and the peak value, passing current that supplies both the load and recharges the

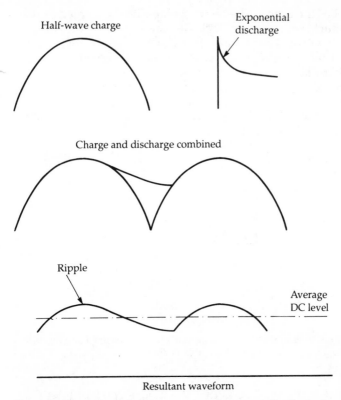

Figure 3.8 Illustrating how the reservoir capacitor supplies the load current for the time during which the diodes are cut off because the input waveform voltage is less than the output level.

reservoir capacitor. For all of the time when the AC output level is below the DC output level and the diodes are cut off, the reservoir capacitor supplies the current, so that the voltage across the reservoir capacitor will drop exponentially, as illustrated in Figure 3.8. The rise and fall of voltage follows part of a sine-wave shape during charging, and part of an exponential discharge at other times, so that its appearance is approximately that of a sawtooth. This wave constitutes the ripple voltage, and most texts assume that it is a sinewave whose RMS value can be given and whose frequency is twice mains frequency for full-wave or biphase half-wave.

These assumptions are not correct, and precise measurements of ripple are by no means simple. In addition, the shape of the

$$\text{Vripple} = \frac{7070 \times I}{C}$$

with I in amps and C in µF

Example: for 1A load current and 5000 µF this predicts a ripple of

$$\frac{7070 \times 1}{5000} = 1.414 \text{ V}$$

Figure 3.9 A very simple analysis of ripple which assumes that the reservoir capacitor is linearly discharged for a time equal to half a cycle.

ripple, though its fundamental frequency is at twice mains frequency for full-wave or bi-phase half wave (mains frequency for a half-wave circuit), contains harmonics which if not suppressed can cause a considerable amount of interference. These harmonics are usually dealt with by other smoothing components, but their presence should be kept in mind; they cannot simply be neglected.

The simplest analysis of ripple is illustrated in Figure 3.9, assuming that the time of discharge is half of the mains cycle time and that the waveform is approximately a sine-wave. As the example illustrates, this leads to a prediction of a 1.4V RMS ripple on a power supply using a 5000µF reservoir capacitor when a current of 1A is drawn. This approach is decidedly on the pessimistic side except for small load currents, the reason being the assumption that the capacitor is discharging for all of a half-cycle whereas in fact the more current is drawn and the lower the output voltage, the less time the capacitor spends discharging.

Another analysis treats the ripple as an AC waveform which will be conducted by the reactance of the capacitor. This results in the formula which is shown in Figure 3.10 and which is on the optimistic side. Neither of these approaches gives correct results at extremes, and they should be used only to give some idea of ripple amplitude for reasonable load currents. For extreme loadings, the output voltage will drop to the amount provided by the

$$\text{Vripple} = \frac{2250 \, I}{C} \text{ with C in µF (for 100 Hz ripple) and I in amps.}$$

Example: for a 5000 µF capacitor, with 1A load current flowing,
ripple = 0.45 V

Figure 3.10 Another formula which treats the problem as one of capacitor reactance, and which gives more optimistic results.

rectifying action itself, as if the capacitor were not present (see earlier), but neither formula deals adequately with this.

Drawing current from a simple rectifier and reservoir capacitor supply will cause a reduction in the DC voltage output and an increase in the ripple amplitude, both caused by the reservoir action alone. In addition, the diode drop and the effect of series resistance in the circuit will cause a further drop in the DC voltage level as load current is supplied. The no-load output voltage will be equal to the AC peak voltage less diode drops, and the on-load voltage will lie somewhere between the peak value and the value for an unsmoothed supply, which is $0.885 \times$ RMS AC value, or $0.632 \times$ peak value for a full-wave rectifier. If we take it that the worst unsmoothed output is 60% of smoothed output for a full-wave bridge or bi-phase half-wave rectifier, then the output value at any practical load value must lie within these limits, ignoring diode drops.

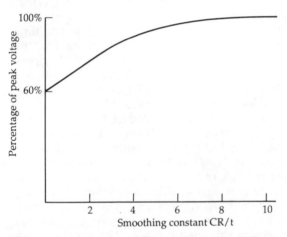

Figure 3.11 A graph which shows approximately the output voltage of a full-wave bridge circuit for a given reservoir capacitor and load time constant.

No really complete analysis is sufficiently easy to use to be of practical interest, but Figure 3.11 shows a graph in which the output voltage is plotted against the time constant of the reservoir capacitor and the load resistance. This, though produced purely by approximations, can be a useful guide to output voltage and its variation. The output voltage is plotted as a percentage of the peak no-load voltage, and the smoothing constant as the fraction CR/t, where C is reservoir capacitance in farads, R is the load resistance in ohms, and t is the time between peaks in seconds — this will usually be 0.01s for a 50Hz supply and full-wave rectification. For a 60Hz supply and full-wave, the time is 0.0083 seconds. Diode drops are ignored.

The other factor that needs to be considered in a power supply of this type is the current rating of the diode. The recharging of the capacitor takes place over a comparatively short time in the cycle, considerably shorter than the time for which the capacitor supplies the load. The amount of current that is drawn in order to charge the capacitor as well as to supply the load for this short time is therefore considerably greater than the amount of average steady current. Fortunately, silicon diodes have large peak-current ratings, and the pulse nature of the current is seldom a problem. The peak current rating of diodes is not always quoted by manufacturers, but is usually more than 10 times the average current rating for a 10ms time. A typical figure quoted for a stud-mounted rectifier 16A diodes is a peak surge of 230A, but this is for a switch-on surge of 10ms, not for repetitive surges.

The internal resistance of a power supply of this type is not a constant, and is not equal simply to the measurable resistances of the transformer windings and the diode. The drop of voltage as load current is taken from the reservoir is the equivalent of an internal resistance, but its value is not constant and not calculable to any degree of accuracy.

Other systems

The bridge-rectifier and capacitor form of supply is almost universally used for electronics purposes, and where regulation is any problem, a stabiliser circuit is added, see Chapter 4. The predominance of the bridge-rectifier and reservoir capacitor type of circuit should not, however, force other systems out of use altogether, and for some applications these options may be more useful.

The single half-wave rectifier is seldom, if ever, used for low-voltage supplies because of its poor regulation and large ripple voltage on load. It finds uses, however, for higher voltage low-current supplies, and in particular for EHT rectification. A simple EHT supply consists of a transformer winding, a diode and a reservoir capacitor, and will be used to supply a current of the order of 1mA. The supply frequency is usually high, not mains, and the reservoir capacitor can be a high-voltage but low-value type. Such capacitors are still available, but have largely disappeared from the catalogues of component suppliers because they are specialised items. The regulation of such a supply is poor, but for many applications this is of little importance. TV EHT supplies are specialised circuits using multipliers (see later) and with stabilisation used.

The bi-phase half-wave circuit is still used, and it has the advantage for low-voltage use that there is only one diode drop in the current path to consider. Against this must be set the higher costs of a transformer with a centre-tapped output and the availability of suitable transformers now that single output windings are so much more common. The circuit is a very useful one if separate stud-mounted diodes are being used, because both diodes can be mounted on the same heatsink without the need for insulation if the studs are connected to the cathodes of the rectifiers.

The little-known alternative to the use of a reservoir capacitor is the use of an inductor as reservoir and smoothing component. Just as a capacitor can store energy in its electric field, an inductor can store energy in its magnetic field, and the inductive form of reservoir offers much better stabilisation of output voltage than its capacitive counterpart. The general trend away from the use of inductive components, however, in the last 30 years has made the use of this type of reservoir system very unusual.

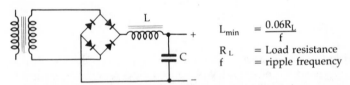

$$L_{min} = \frac{0.06R_L}{f}$$

R_L = Load resistance

f = ripple frequency

Figure 3.12 Inductive input filter circuit, which uses an inductor to store energy rather than a capacitor. The output level is lower than the corresponding capacitor reservoir circuit, but the stabilisation is much superior.

The basic circuit and its requirements are illustrated in Figure 3.12 with a bridge rectifier feeding the inductor. The calculation for minimum inductance is, as always, approximate, and the value obtained from this calculation should be regarded as an absolute minimum — a realistic minimum would be 25% greater than the predicted value, and for a practical power supply an inductance value of several times this minimum would be used. The ripple content is determined by the value of capacitance following the inductor and is expressed in terms of the LC combination. Care should be taken to avoid resonance of this circuit with mains frequency or double mains frequency, as this would cause excessive currents that could burn out the diodes. For a full-wave rectifier using 50Hz or 60Hz mains, this amounts to avoiding any value of L × C that gives a figure anywhere near 2.5×10^{-6}, for 50Hz mains, or 1.7×10^{-6} for 60Hz mains. (with L in henries and C in farads). In more practical units of henries and microfarads, this means avoiding a figure of 2.5HμF (50Hz) or 1.7HμF (60Hz).

The off-load output voltage for this type of rectifier-reservoir combination is considerably lower, about 64% of peak AC voltage for a full-wave bridge circuit, but since this is much the same as is achieved without smoothing, the regulation is considerably better. The DC current output when on load is about 94% of AC current from the transformer. The drawback is the use of a large and heavy inductor, particularly for low-voltage high-current supplies, because the resistance of the inductor will make the internal resistance of the supply much larger than that of a comparable capacitor-reservoir type, and this reduces the advantages of better inherent regulation.

A variation on the inductance reservoir system is the use of a current-variable inductor or *swinging choke*. This can be a physically smaller and lighter component, whose minimum inductance (and resistance) can be considerably less than that used for a conventional inductor-smoothed circuit. Since the system is never used for low-voltage supplies, swinging-chokes are by now almost unobtainable.

Voltage multipliers

A voltage multiplier circuit is one which provides a DC output whose level is higher than the peak AC voltage from the transformer or other AC source. The simplest form of circuit, the half-wave doubler, is illustrated in Figure 3.13. If we imagine this

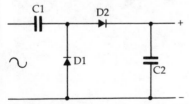

Figure 3.13 The half-wave voltage doubler circuit which is the simplest form of multiplier circuit.

at the point of being switched on with all capacitors uncharged and no load, then the first positive half cycle will cause D2 to conduct, charging C2 to the peak of the AC voltage. When the AC voltage reverses, D1 will conduct, and this will now charge C1 to the peak voltage, positive at the cathode of D1. The next positive cycle of AC will now be superimposed on to the existing voltage at the cathode of D1, equal to peak voltage, so that on the next positive cycle, C2 will be charged to twice the peak voltage of the AC. This capacitor must therefore be rated for at least twice AC peak voltage working. From now on, the voltages that are maintained will be equal to peak AC at the cathode of D1 and twice peak AC at the cathode of D2.

The regulation of a voltage doubler is very poor, mainly because the capacitors have a fairly large reactance even at the higher supply frequencies which are almost always used along with multiplier circuits. Unlike the rectifier circuits for low-voltage supplies, there is no real lower limit to the output of a multiplier circuit on heavy load, and if the load resistance fluctuates considerably some stabilisation will be needed. Most multiplier circuits make use of waveforms derived from oscillators, and a control circuit for stabilisation samples the output voltage and uses this to control the amplitude of the oscillator. As for any other rectifier-reservoir type of supply, the ripple amplitude increases as the DC output level drops.

Figure 3.14 shows an alternative form of doubler, in which one lead from the transformer is connected to the junction of two capacitors. This configuration has the advantage that each capacitor is subjected to only the AC peak voltage unlike the second capacitor in the circuit of Figure 3.13. The AC input is floating in this design, which can make insulation requirements easier, and the diodes also share the peak reverse voltage.

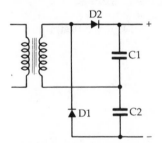

Figure 3.14 Another form of doubler circuit in which neither side of the AC input is earthed.

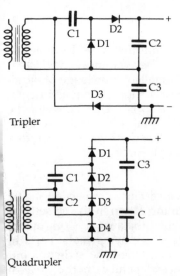

Tripler

Quadrupler

Figure 3.15 Voltage tripler and quadrupler circuits.

Voltage tripler and quadrupler circuits, Figure 3.15, can make use of the same principles of using a capacitor to charge by using current from one diode and then to superimpose this voltage on to another conducting diode along with an AC half-wave. The more the number of stages of multiplication, the poorer the regulation, but for load currents which in many cases do not exceed 1mA this is usually unimportant. For TV receiver use, voltage multipliers make use of the pulse waveforms in the line-output transformer, and the use of miniature EHT silicon diodes allows a combination of winding and rectifier, Figure 3.16, to be used, avoiding the need for high-voltage capacitors. The capacitor that is shown in the diagram is in fact the stray capacitance of the carbon coating on the outside of the picture tube.

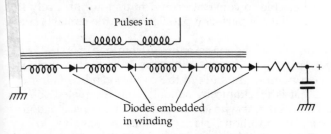

Figure 3.16 A form of voltage multiplier which makes use of diodes incorporated into the winding of a pulse transformer, with no need for capacitors.

Switches and fuses

Switches and fuses are not usually considered as forming an important part of a power supply, but they cannot be neglected, particularly as so few users have any detailed knowledge of either type of component. In addition, if a power supply is not an integral part of a circuit, connections will have to be made between the power supply and the circuit which it supplies. These connections will usually be by way of plugs and sockets which therefore become another part of the power-supply circuit.

Switches

Switches are required to make a low-resistance connection in the *on* setting, and a very high-resistance insulation in the *off* setting. The resistance of the switch circuit when the switch is on (made) is determined by the switch contacts, the moving metal parts in each part of the circuit which will touch when the switch is on. The amount of the contact resistance depends on the area of contact, the contact material, the amount of force that presses the contacts together, and also in the way that this force has been applied. For a power supply, the main consideration in switch contacts is the surge of current as contacts are made (due to charging of capacitors) and any reverse voltage surge that occurs when the switch is opened (because of decreasing current in the transformer).

If the contacts are scraped against each other in a wiping action as they are forced together, then the contact resistance can often be much lower that can be achieved when the same force is used simply to push the contacts straight together. In general, large contact areas are used only for high-current operation and the contact areas for low-current switches as used for electronics circuits will be small. The actual area of electrical connection will not be the same as the physical area of the contacts, because it is generally not possible to construct contacts that are precisely flat or with surfaces that are perfectly parallel when the contacts come together.

A switch contact can be made entirely from one material, or it can use electroplating to deposit a more suitable contact material. By using electroplating, the bulk of the contact can be made from any material that is mechanically suitable, and the plated coating will provide the material whose resistivity and chemical action is more suitable. In addition, plating makes it possible to use materials such as gold and platinum which would make the switch impossibly expensive if used as the bulk material for the contacts.

It is normal, then, to find that contacts for switches are constructed from steel or from nickel alloys, with a coating of material that will supply the necessary electrical and chemical properties for the contact area. The usual choice of material for mains switches for power supplies is silver or silver-plated nickel. Note that the ratings for such switches always assume a resistive load, and derating should be applied for an inductive load.

Ratings
Switch ratings are always quoted separately for AC and for DC, with the AC rating often allowing higher current and voltage limits, particularly for inductive circuits. If switches are used on the DC lines of a power supply, some care over specifying the switches will be needed. When DC through an inductor is decreased, a reverse voltage is induced across the inductor, and the size of this voltage is equal to inductance multiplied by rate of change of current. The effect of breaking the inductive circuit is a pulse of voltage, and the peak of the pulse can be very large, so that arcing is almost certain when an inductive circuit is broken unless some form of arc suppression is used.

Arcing
Arcing is one of the most serious of the effects that reduce the life of a switch. During the time of an arc very high temperatures can be reached both in the air and on the metal of the contacts, causing the metal of the contacts to vaporise, and be carried from one contact to the other. This effect is very much more serious when the contacts carry DC, because the metal vapour will also be ionised, and the charged particles will always be carried in one direction. Arcing is almost imperceptible if the circuits that are being switched run at low voltage and contain no inductors, because a comparatively high voltage is needed to start an arc. For this reason, then, arcing is not a significant problem for switches that control low voltage, such as the 5V or 9V DC that is used as a supply for solid-state circuitry, with no appreciable inductance in the circuit. Even low voltage circuits, however, will present arcing problems if they contain inductive components, and these include relays and electric motors as well as chokes. Circuits in which voltages above about 50 V are switched, and particularly if inductive components are present, are the most susceptible to arcing problems, and some consideration should be given to selecting suitably rated switches, and to arc suppression, if appropriate.

Temperature range

The normal temperature range for switches is typically −20°C to +80°C, with some rated at −50°C to +100°C. This range is greater than is allowed for most other electronic components, and reflects the fact that switches usually have to withstand considerably harsher environmental conditions than other components. The effect of very low temperatures is due to the effect on the materials of the switch. If the mechanical action of a switch requires any form of lubricant, then that lubricant is likely to freeze at very low temperatures. Since lubrication is not usually an essential part of switch maintenance, the effect of low temperature is more likely to be to alter the physical form of materials such as low-friction plastics and even contact metals.

Flameproof switches must be specified wherever flammable gas can exist in the environment, such as in mines, in chemical stores, and in processing plants that make use of flammable solvents. Such switches are sealed in such a way that sparking at the contacts can have no effect on the atmosphere outside the switch. This makes the preferred type of mechanism the push-on, push-off type, since the push button can have a small movement and can be completely encased along with the rest of the switch.

Switch connections can be made by soldering, welding, crimping or by various connectors or other plug-in fittings. The use of soldering is now comparatively rare, because unless the switch is mounted on a PCB which can be dip-soldered, this will require manual assembly at this point. Welded connections are used where robot welders are employed for other connection work, or where military assembly standards insist on the greater reliability of welding. By far the most common connection method for panel switches, as distinct from PCB mounted switches, is crimping, because this is very much better adapted for production use. Where printed circuit boards are prepared with leads for fitting into various housings, the leads will often be fitted with bullet or blade crimped-on connectors so that switch connections can be made.

Fuses

All electronic circuits, unless of the microwatt variety powered by a high-impedance battery, must be fused so as to prevent damage that would be caused by excess current. The choice of fuses from the usual bewildering variety is much more strictly governed than the choice of other hardware items, however, because there are often national regulations which must be obeyed if fuses are being replaced or if electronic equipment is being exported. European standards specify fuses of 20mm length and 5mm diameter (20 × 5 fuses), whose specification is outlined by IEC 127. For the USA and Canada, however, the fuse standards are UL 198G and CSA22-2 No. 59 respectively, using 1.25 inch × 0.25 inch fuses whose characteristics in terms of blowing time and current are quite different from the European standards. These fuses are not of interchangeable dimensions but nevertheless great care should be taken not to mix the types. This is particularly important in the UK where 1.25 inch fuses which are to the IEC 127 standards on current and time characteristics are in use along with the 20 mm type.

Rating

The aim of a fuse is to interrupt current in the event of a fault that causes excessive current to flow. The subject is not nearly as simple as this would suggest, however, and if you thought that a 1A fuse would blow when the current reached a value of 1A then you have not been heavily involved in choosing fuses. If a fuse is used in a circuit in which a high voltage would exist after the fuse blows, then that voltage placed across the fuse might be enough to cause an arc-over, so that current was not interrupted when the fuse blew. Fuses carry a voltage rating, and should not be used beyond that rating, so that a fuse rated at 125V should not be used in a circuit in which 240V could exist across a blown fuse. A fuse can be safely used at voltages lower than the rated maximum, but not at any voltage higher than the stated maximum.

The interrupt test rating of a fuse is another value which is not well known. A fuse rated at 1A might never have to bear a current of 1000A or more, but its behaviour at such high currents is important. If the fuse continued to pass current, for example, because of conduction by the metal vapour from the wire, it would once again be unsatisfactory. Approval testing for fuses is therefore carried out at very high currents as well as with high voltages across the fuse, and this figure is often quoted.

The quantity that is always quoted for a fuse is the nominal current rating, but precisely what that means depends on the standard to which the fuse has been constructed. A fuse which is rated at 1A, for example, will not necessarily blow at a current of 1A, because the blowing of a fuse is a complicated process that involves current and time. The current determines the dissipation in watts, but the rate at which the fuse heats is determined by the material, the surroundings, and the amount of conduction of heat from the connections. The blowing of the fuse is also affected by the melting-point temperature of the metal of the fuse, so that the construction of a fuse affects its blowing current more than any simple theoretical considerations. Since the behaviour of a fuse is so complex, various standards exist to provide guidelines on the current-time limits. The three main fuse standards that you are likely to come across are the UL 198G (USA), CSA22.2 (Canada) and the IEC 127 (Europe, including the UK). These standards are substantially different; a European 1A fuse would be rated in the USA as 1.35A, and this is general. European fuse ratings are aimed at protection against large overloads caused by short circuits, fuse ratings in the USA aim to provide overload protection from currents that are rather less than short-circuit levels.

Very great care must therefore be taken with equipment of overseas origin, or with equipment intended for export, that the fuse types are correct as well as the ratings. Probably the easiest way to remember the equivalence of the different standards of fuses is that the fast fuses are rated in the ratio 10/9 (10A US fuse = 9A Eurofuse) and for the others the ratio is 8/6 (8A US fuse = 6A Eurofuse).

Fuses are grouped in five major categories according to their current-time characteristics. At one end of the scale, semiconductor circuits need fuses that will act very quickly on short-circuit conditions. These are now described as super quick-acting fuses, coded FF. These, at 10 times rated current, will blow in a millisecond or less, and for twice the rated current the blowing time will be 50ms or less. The next group comprises the quick-acting Class F fuses (classed as *Normal Blo* in the USA), used for general-purpose protection where current surges are unlikely to be encountered. These have a slower blowing characteristic, some 10ms for ten times rated current and just over 100ms for twice rated current.

The medium time-lag fuses, type M, will withstand small current overloads that might be caused by charging capacitors. These fuses will blow after about 30 ms on a ten-times current overload, and after about 20 seconds on a two-times overload. The long time for the smaller overload allows for current surges that are not particularly large but of quite long duration. The Time-lag type T fuses (classed as *Slo-Blo* in the USA) will blow in 100ms for a ten times overload and in about 20 seconds for a twofold overload. Super time-lag class TT fuses allow for 150 ms at a tenfold overload and 100 seconds at a twofold overload.

All fuse ratings are measured at 20°–25° ambient temperature, and because the blowing of a fuse is a thermal process, the ratings of a fuse are affected by changes in the ambient temperature. The slower-blowing fuses in particular, which depend on the use of some heatsinking to delay blowing, need to be derated if they are to be used at high ambient temperatures, and for these types, derating to 60% of nominal value is recommended if the fuse is to be used at 100°C. The derating for the faster-blowing types is considerably less, but in general, fuses should not be located at points in a circuit where high temperatures exist (e.g. next to a set of power-transistors) unless this is done deliberately as a safety measure. Quite irrespective of any derating due to ambient temperature, the normal current through a fuse can be the fuse rated current for a European type of fuse, but not more than 75% of rated current for a USA/Canada type of fuse.

Fuses are resistors, and though the larger capacity fuses have a negligible resistance, this is not true of the smaller types. Fuse resistance is not usually quoted by suppliers, but it can add to the internal resistance of a power supply and upset stabilisation to some extent, though for the larger rated fuses the contact between fuse and fuseholder contributes more resistance in some cases.

Fuse specifications provide a table of currents, in terms of rated current, along with maximum and minimum times for which the fuse should withstand such currents. The USA/Canada specification, for example, provides that a fuse should be able to withstand a current of twice the rated value for more than 5 seconds, and the IEC ratings provide for minimum times for which a fuse should continue to conduct at small overloads, along with both maximum and minimum times for blowing on larger overloads up to 10 times rated current. In addition to the normal standard 20mm or 1.25 inch fuses, there are fuses both in these sizes and in miniature sizes which have very different current-time characteristics.

Fuses that are used in power supplies for semiconductor circuits will offer no protection against damaging transients that are of much shorter duration than would blow a fuse. For dealing with type of transient, a crowbar circuit is required, and this type of circuit is dealt with in Chapter 4.

Connectors

Plugs and sockets present the same problems of contact resistance as switches, but it can usually be assumed that the connectors will never be connected nor disconnected while voltages are present. The important points here are to observe safety standards for mains connectors, using either a clamped cable into the power supply, or the BS 4491 type of plug and socket connector. There are many varieties of connectors that are rated for fairly high voltage and current use, but which are not suitable for use with mains connections, usually because they do not comply fully with safety standards. Never use any form of unapproved connector for mains wiring.

For power supply outputs, the choice is wider, but preference should be given to connectors that are designed for DC as distinct from audio, video or RF waveforms. Some audio connectors have very small areas of contact and will cause problems, mainly of varying resistance, if used with DC supplies. Some of the most suitable connectors for power supplies that provide several voltage levels are the connectors for professional video circuits.

4 Stabilisation

Stabilisation means the maintenance of a fixed level of voltage or (less commonly) current from a power supply. Though stabilisation is usually nowadays associated with IC stabilisers, this is just one method out of many, and some other methods have peculiar advantages for special purposes. In order to make effective use of stabilisation (or regulation) methods, it is essential to understand the principles involved in conjunction with what has already been mentioned in connection with power supplies.

To start with, a stabiliser operates by wasting power, and it is usually needed to stabilise against two causes of variation of output. The first is alteration of the mains supply, caused by sudden loads (lifts, electric kettles being switched on after the end of popular TV programmes etc.). A good stabiliser should not pass any of these voltage surges to the DC output, and in some cases more than simply stabilising is needed. The other effect is the variation of load current caused by the internal resistance of the power supply, including the effect of the reservoir capacitor. A stabiliser should be able to keep the output terminal voltage steady despite current variations between full rated load and zero load.

The peak voltage output from a power supply will fall at full rated output current to some lower value, and this lower value is the one which can be stabilised. In fact, to allow for the effect of the stabiliser itself, the output voltage will be lower still. This implies that the amount of waste can be considerable. For example, a 5V DC output from a stabiliser will require an input to the stabiliser of about 6V absolute minimum, and in order to provide this 6V minimum at full current, the off-load voltage may be nearer 8–10V, depending on current. The ripple that is present should also be removed by the stabiliser, and this requires that the negative peak of ripple voltage should not take the output of the rectifier-reservoir system below the minimum voltage for the stabiliser. As ripple is usually quoted in RMS terms, it is not always easy to find out what the minimum transient voltage output from the reservoir capacitor will be.

All of this leads to the use of fairly generous voltage levels at the input to the stabiliser, but this in turn can present dissipation problems. IC stabilisers will have a maximum input voltage stipulated, and this can be surprisingly high, as high as 35V for a 5V stabiliser. The voltage difference exists across the stabiliser, however, and the stabiliser will have to dissipate power equal to the voltage excess multiplied by current flowing. For example, if a supply into a 5V stabiliser is 15V at a current of 1A, then the excess voltage is 10V and the power dissipated is 10W. Calculations of dissipation, and use of heatsinks are both dealt with later in this chapter.

Stabilisation principles

Stabilisation depends on the use of a non-linear circuit or component placed in series with the current or in parallel with the voltage of a power supply. The basic circuits are illustrated in Figure 4.1, showing a non-linear resistor either in series or in

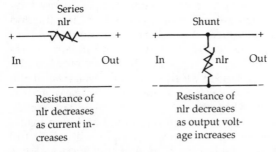

Figure 4.1 The most basic series and shunt stabiliser circuits, making use of a non-linear resistor.

shunt. The principle of the series stabiliser is that the resistance of the non-linear component becomes less as current increases, so compensating for dropping voltage of the reservoir by having a lower drop across this resistance. The shunt element operates by passing a current which alters inversely with the load current, so that a drop in the voltage level causes a disproportionate drop in the current through the stabiliser. Stabilisation systems using PTC thermistors or non-linear resistors are quite widely used commercially, particularly for circuits in which the voltage or current levels make the use of conventional stabilisers impossible.

Another form of stabilisation operates through the supply, controlling the AC input to a rectifier by way of thyristors or other methods, with the control being derived from a comparator circuit which compares the output DC level to that of a stable source. This type of stabilisation is also quite common in commercial circuits, and its most usual form is as part of a switch-mode supply, dealt with in Chapter 5.

Stabilisation with Zener diodes

All stabilisation requires some form of comparison between a variable voltage and a steady standard voltage, and the usual form of standard is provided by the Zener diode. The Zener diode, named after the discoverer of its principles, Clarence Zener, is not quite what it seems, because two effects are present, one called avalanche breakdown and the other called Zener breakdown. For most purposes, the differences are mainly of interest to researchers, but when we come to look at temperature effects, the differences become more important.

Breakdown voltage

The important feature of the Zener diode is that it is intended to be used with reverse bias, although it possesses a normal forward characteristic like any other silicon diode. The normal reverse characteristic of any silicon diode is to withstand reverse voltages until a threshold value at which reverse current increases rapidly. The Zener type of diode is constructed so that the breakdown of resistance for reverse voltage is much more rapid, leading to a characteristic of the type illustrated in Figure 4.2, with a very steep slope at the breakdown voltage. This implies that over a large range of reverse currents, the reverse voltage will be almost constant, so that this type of diode can be used as a constant-voltage source.

The voltage for breakdown is, for a normal silicon diode, made as high as possible subject to the requirements of the diode, but for a Zener diode, altering the doping of the silicon allows diodes to be made with a range of breakdown voltage levels. Diodes are usually selected so as to offer a set of breakdown voltage levels that follows the normal tolerance sequence, and using values that are written following the convention of BS 1852. This uses the letter

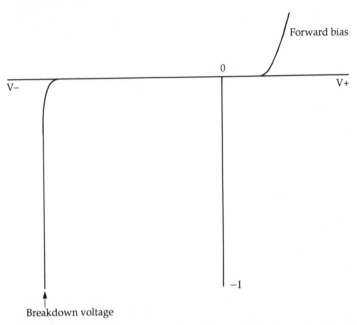

Figure 4.2 The characteristic of a Zener diode, in which the reverse characteristic is the point of interest.

V in place of the decimal point so as to avoid any confusion due to misplaced points or to the use of a comma in place of a point, so that you will see Zener diode voltages quoted in terms of numbers such as 4V7, 5V6, 6V8, 7V5 and so on. The effectiveness of a Zener diode for providing a stable output reference voltage is measured by its slope resistance, the equivalent of internal resistance. This is not a constant, but something that depends on doping levels, and is not the same for diodes of different power dissipation ratings. For example, its value is smallest for diodes of 0.5W dissipation whose doping corresponds to a voltage break-down level of between 6V and 8V. The slope resistance value, shown plotted against breakdown voltage in Figure 4.3 for 0.5W diodes, is due to two conflicting effects which are in balance with each other when this level of doping is used.

A typical circuit for voltage reference is illustrated in Figure 4.4. The Zener diode is connected in series with a resistor which is used to limit the current. If the supply voltage input varies, but does not fall as low as the Zener voltage, then the current through the

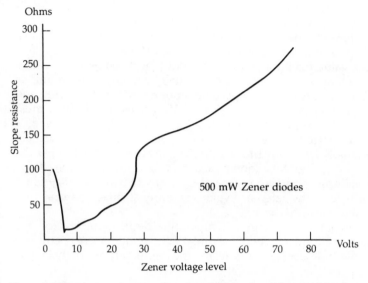

Figure 4.3 The slope resistance value plotted against breakdown voltage for a range of Zener diodes shows a minimum value at around 6 – 8V.

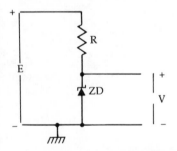

Figure 4.4 Using a Zener diode in a voltage reference circuit. The resistor limits the amount of current through the diode, and the supply voltage must remain above the Zener breakdown voltage for the stabilisation to be effective.

diode will vary but the voltage across the diode will be almost constant. The current through the resistor is given by:

$$\frac{E-V}{R}$$

where E is the supply voltage and V is the Zener breakdown voltage.

For example, if the Zener breakdown level is 7.5V and there is a 12V supply available, the first step in finding the required resistor is to determine the current level required through the Zener. Most suppliers will quote maximum permitted Zener current, but few quote a value for the minimum current that is required to sustain Zener action. A current of 1mA is usually a realistic maximum, so that 2 mA is a reasonable specified amount. For 2mA flowing and a voltage difference of $12 - 7.5 = 4.5$V, the resistance value must be $4.5/2 = 2.25$K, and in practice a 2K2 resistor would be used.

If the current level drops to a very low value, usually considerably less than 1mA, or the voltage level drops to near the Zener level, the breakdown action will cease and the diode becomes non-conducting. This also means that the voltage is no longer held at the reference value, and any stabilisation based on this voltage will cease.

The breakdown voltage is temperature dependent, but for one voltage breakdown level of around 4V the temperature coefficient is almost zero. Below this level, the temperature coefficient is

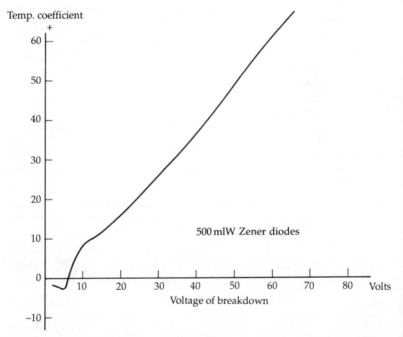

Figure 4.5 Temperature coefficient values plotted against Zener breakdown voltage for a range of Zener diodes.

negative, and above this level the temperature coefficient is positive. Figure 4.5 shows a graph of temperature coefficient values plotted against voltage breakdown levels. This effect arises because the observed temperature coefficient is the sum of two separate temperature effects which act in opposite directions and cancel each other out for a breakdown voltage of around 4V. The effect of temperature coefficient is to change the voltage breakdown level as the temperature changes. For Zener diodes, the temperature coefficient is usually quoted as a percentage change per °C, so that a typical value for a 7V5 Zener is +0.065%/°C. Using this as an example, if the normal running temperature is 25°C and the Zener is used at 75°C, then the temperature difference is 50°C and this will cause a change in voltage of 0.065% × 50 = +3.25%. For a 7V5 Zener, a change of +3.5% is 7.5 × 3.5/100 = +0.2625V, making the voltage about 7.76. If the temperature coefficient is negative, as it is for values below 4V0, then the change is also negative, meaning that the voltage level will drop as the temperature is raised.

Zener diodes are supplied in a range of breakdown voltages, typically 2V7 to 62V, and in a large variety of power ratings. Small Zeners are rated at around 0.5 to 1.3W, subject to a derating of typically 9mW per degree above a temperature of 25°C. Zeners of 20W rated dissipation are available, but there is usually little point in using large wattage Zener diodes in simple stabiliser circuits, because voltage stabilisation can be accomplished more cheaply and efficiently at the same power levels by using a stabiliser IC.

The more specialised forms of Zener diodes have been developed for very small temperature coefficient or for very stable voltage control. Temperature-compensated diodes can achieve a temperature coefficient as low as 0.001% per °C, and voltage reference diodes can supply a precise voltage which is very stable against temperature or current changes, and also with a very low noise output. The most precise and stable voltage reference diodes, however, are bandgap diodes which make use of the forward voltage across junctions to give a low-voltage reference source of 1.2. Bandgap diodes are available as discrete components, but are more often incorporated as part of an IC in order to provide a stable reference voltage.

Zener circuits

The simple Zener circuit of Figure 4.4 can be used as a stabiliser in its own right for small-scale circuits, with an adjustment of the value of series resistance. Suppose that the load is to be stabilised at 7.5V for a current range in the load from 0 to 10mA. The stabiliser must be arranged so that the current through the Zener when the load is cut-off will be greater than the maximum load current. In this way, when the load current alters, it will always be balanced by an opposing change of current through the Zener, with the voltage remaining constant. Using the example in Figure 4.6, if the minimum Zener current is to be 2mA, then allowing for

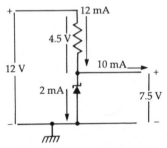

Figure 4.6 Using the Zener diode circuit as a simple shunt stabiliser. The current that flows with no load applied must be greater than the maximum load current.

10mA through the load means that with the load cut off, the Zener must pass 12mA. If the supply voltage at a current of 12mA is 12V, then this allows 4.5V drop across the resistor for a current of 12mA, making the resistor value equal to 4.5/12 = 0.375K, or 375Ω. In practice, a 330Ω resistor would be used. This resistor will pass a constant 12mA, so that the dissipation of the resistor will be 4.5 × 12mW = 54mW, well within the capabilities of a small resistor. The dissipation of the Zener when 12mA is passing will be 7.5 × 12mW, which is 90mW, once again quite small. Note that the dissipation from the Zener is maximum when the load is not drawing current and the dissipation from the resistor is constant whether the load is drawing current or not.

The normal shunt-stabilising action of a Zener diode can be assisted by using a transistor controlled by the Zener diode as the shunt element. An example of this type of action is illustrated in Figure 4.7, in which a PNP transistor is connected as an emitter-follower with the base voltage determined by the voltage level of the Zener diode. This will cause the emitter voltage also to be reasonably well stabilised at a level which in this example is 0.6V higher than the voltage of the Zener diode. This type of circuit is

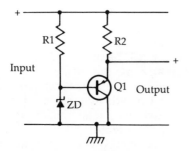

Figure 4.7 A shunt stabiliser which makes use of a transistor and a Zener diode. This arrangement is sometimes called an 'amplified Zener' circuit.

sometimes called an 'amplified Zener' circuit, but it is seldom used nowadays because of the predominance of IC stabilisers.

The Zener used in this way, amplified or not, is acting as a shunt stabiliser, but this form of stabilisation is inefficient, with a substantial amount of dissipation in the resistor whether current is being drawn by the load or not. The series stabiliser is more effective, and is the method that is used by IC stabilisers. Though it is most unlikely that anyone now would construct a series stabiliser in preference to using an IC type, it is important to know the circuit and the principles involved, because this leads to better appreciation of how to make use of the IC types.

The most basic form of series stabiliser is illustrated in Figure 4.8. The Zener diode is connected through a resistor to the unstabilised supply voltage, and the diode voltage is applied to the base of the power transistor. The resistor from the output of the stabilised supply at the emitter of the transistor is not essential, but if no other load is connected it helps to establish correct voltage

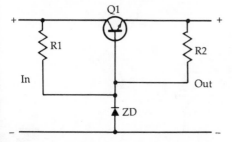

Figure 4.8 An elementary series stabiliser circuit with one transistor. This is also a useful smoothing circuit if the transistor has a high h_{fe} value.

levels quickly when the circuit is switched on. The output of this circuit is not identical to the Zener output, and the variation is rather greater. The Zener diode establishes the base voltage for the transistor, but the voltage at the emitter will be about 0.6V less, and the difference will become greater as the load current increases. The transistor will usually be mounted on a heatsink, and the current to the Zener diode must be adequate to provide for the base current that the transistor will need when maximum load current is passing between collector and emitter.

In this type of circuit, the transistor dissipates power only as and when the load dissipates power, unlike the shunt form of stabiliser. The load is stabilised against both input voltage variations and load current variations, though the stabilisation against load current variation is only as much as can be provided by the use of the transistor in this emitter-follower circuit. The maximum variation of voltage will be equal to the variation of base to emitter voltage for the range of current that is being used. This may amount to no more than about 0.2V.

The stabilisation of the simple series circuit may be adequate for some purposes, but for a closer control of voltage, a comparator amplifier method must be used. The principle, as distinct from a practical circuit, is illustrated in Figure 4.9. The Zener diode, supplied from the unstabilised voltage, is used to provide a constant voltage to one input of the comparator. The other input is supplied by way of a voltage divider from the output at the emitter of the power transistor. The signal from the comparator is used to control the base voltage and current of the power transistor so that a rise in output voltage will cause the base voltage to be

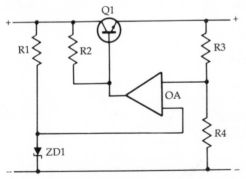

Figure 4.9 Using a comparator amplifier along with a Zener-stabilised reference and a sampling potential divider, controlling a series transistor.

reduced and a fall in output voltage will cause the base voltage to be increased. Because of the amplification in the comparator, providing a large loop gain in a negative feedback circuit, the control of the output voltage level can be much closer. This is the basis of the type of circuit that is used in IC stabilisers.

IC stabilisers

Since it became possible to construct voltage stabilisers in IC form and make them work reliably (something that took a little longer to achieve), the construction of a stabilised supply from discrete components has become almost a lost art. It is almost impossible to think of any good reason for not using an IC stabiliser if one is available for the range of voltage and current that is required, and since there is a stabiliser suitable for virtually every application in low-voltage electronics, discrete circuits are now very rare.

This does not mean, however, that all voltage stabilisation problems can be solved by connecting an IC stabiliser into an existing power supply, and some appreciation of the action of the circuitry helps considerably in understanding what is involved in the use of the IC. In particular, since the IC incorporates power transistor, comparator, Zener diode and resistors into one unit, the Zener will be subject to higher temperatures than would usually be encountered in a discrete circuit. The use of heatsinks and calculations of temperature will therefore form an important part of the design of a power supply that uses this type of stabiliser.

IC regulators are available in a bewildering variety of types, but the simplest are the familiar fixed voltage types such as the 7805. The final two digits of this type number indicate the stabilised output, 5V for this example. The 7805 is a three-pin regulator which requires a minimum voltage input of 7.5V to sustain stabilisation, with an absolute maximum input voltage of 35V. The maximum load current is 1A and the regulation against input changes is typically 3–7mV for a variation of input between 7.1V and 25V. The regulation against load changes is of the order of 10mV for a change between 5mA and 1.5A load current. The noise voltage in the band from 10Hz to 100kHz is 40–50µV, and the ripple rejection is around 70db. Maximum junction temperature is 125°C, and the thermal resistance from junction to case is 5°C/W.

This stabiliser, used extensively for power supplies in digital equipment provides a useful source of examples that illustrate the way that stabilised supplies can be designed around an IC. The

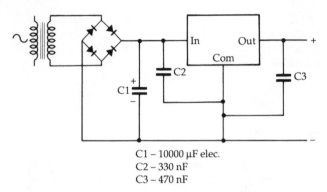

C1 – 10000 µF elec.
C2 – 330 nF
C3 – 470 nF

Figure 4.10 A recommended circuit for a power supply using an IC regulator. The physical positions of the capacitors are very important.

recommended circuit is shown in Figure 4.10 with the diode bridge and reservoir capacitor. The capacitors that are shown connected each side of the IC are very important for suppressing oscillations and must not be omitted. In particular, the 330n capacitor at the input must be wired across the shortest possible path at the pins of the IC. The maximum allowable dissipation is 20W, assuming an infinite heatsink, and the actual capabilities are determined by the amount of heatsinking that is used. If no heatsink is used, the thermal resistance of the 7805 is about 50°C/W, and for a maximum junction temperature of 150°C this gives an absolute limit of about 2.5W, which would allow only 2.5V across the IC at rated current, an amount only just above the absolute minimum.

The thermal resistance, junction to case, is 4°C/W, and for most purposes, the IC would be mounted on to a 4°C/W heatsink, making the total thermal resistance 8°C/W. This would permit a dissipation of about 15.6W, which allows up to 15V or so to be across the IC at the rated 1A current. This is considerably more useful, since a 5V supply will generally be provided from a 9V transformer winding whose peak voltage is 12.6V, making it impossible to cause over-dissipation at the rated 1A current since the voltage output from the reservoir capacitor will be well below 12.6V when 1A is being drawn. Figure 4.11 shows the physical form of the recommended heatsink and the two forms of the casing of the IC itself.

The 78 series of stabilisers are complemented by the 79 series, which are intended for stabilisation of negative voltages. The circuits that can be used are identical apart from the polarity of

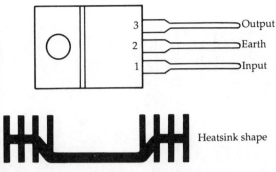

Figure 4.11 The shape of the 78xx type of stabiliser IC and a suitable heatsink.

diodes and electrolytic capacitors, and the range of currents is substantially the same as for the 78 series.

Altering voltage output levels

The voltage output from a stabiliser of fixed voltage can be raised only by increasing the voltage level at the common pin, and if this is done two precautions are required. One is that the unstabilised input voltage must be adequate to cope with the new output level — you cannot expect a 9V transformer winding which has been used for a 5V stabilised supply to be able to supply an output of 9V at 1A. The other point is that there is a current of about 4.5mA to earth through the common lead of the IC, and this itself will contribute to the raising of the output voltage if a resistor is connected between the common lead and earth. Figure 4.12 shows the circuit in which the resistors R1 and R2 are used, with current I being provided by the IC. The voltage output of this arrangement is:

$$V = IR_2 + V_r(1 + \frac{R_2}{R_1})$$

with the current through R1 at least five times the current from the stabiliser common lead, implying a current of at least 22.5mA. Resistor values are assumed to be in kilohms.

The formula as it stands makes it possible to predict what voltage output will be obtained for given values of R1 and R2, but it is not particularly useful for finding what values of R1 and R2 to use for

83

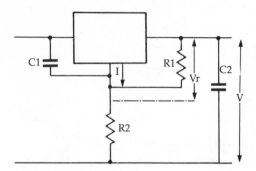

Figure 4.12 Raising the stabilised output voltage level of a fixed-voltage stabiliser by means of two resistors. The effect of the current flowing from the common lead of the stabiliser must be taken into account.

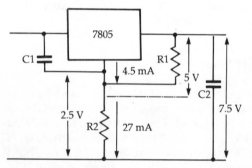

Figure 4.13 A practical circuit using a fixed-voltage stabiliser at a higher output level, showing the currents and voltages to be calculated.

a given voltage output. For this latter application, it is better to work out the resistor values from the known currents and voltages in the circuit, and an example will be considerably more useful than a formula.

Suppose that we want to make a 5V regulator operate at 7.5V, using a 7805 type of regulator so that the current to earth from the common lead is 4.5mA. The circuit will be as shown in Figure 4.13, with the voltage and current levels, remembering that R1 must carry a current that is at least five times as much as the 4.5mA of the regulator. The voltage and current values are then as shown also in Figure 4.13. The voltage across R2 is 2.5V for a current of 27mA, so that the value of R2 is 92.6Ω. The voltage across R1 is 5V for a current of 22.5mA, so that the value of R1 is 222Ω. The output is correct for these values, which could be obtained by using shunt and series combinations of resistors, but it is useful to go back to the original formula to find how far out the results would be if we took the nearest standard values of 220R and 100R. These would give:

$$V = 5(1 + 100/220) + 4.5 \times 0.1$$

which is 7.72V. If the stabilisation at this level were more important than the actual value, the use of the standard components would be acceptable. An alternative is to juggle the values, knowing that these values give a result that is on the high side. By using the 5% tolerance value of 91R for R2 and 220R for R1, however, the value of 7.47V can be obtained, and this must be as close as anyone is likely to need.

The regulation of the circuit of Figure 4.12 is not as good as that of the stabiliser IC itself, because of the way that the division ratio of the resistors is affected by the current from the common lead of the IC. The regulation of a non-standard voltage using a fixed-voltage stabiliser can be greatly improved by using an emitter follower circuit to establish the voltage at the common connection. The circuit is illustrated in Figure 4.14, using a PNP transistor

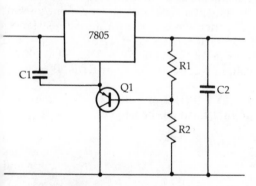

Figure 4.14 A better method of using a fixed-voltage stabiliser to provide higher voltage levels, using an auxiliary transistor.

whose emitter is connected to the common return pin of the stabiliser IC.

The calculation is simpler in this case, but the base-emitter voltage of the PNP transistor has to be taken into account. The voltage at the junction of R1 and R2 is found from the usual potential divider formula, and this voltage, added to the regulator voltage plus 0.6 V gives the output voltage. Put into a formula, this becomes:

$$V_{out} \frac{(V_r + 0.6) \times (R1 + R2)}{R1}$$

For the more usual problem of calculating the resistor values from the required voltage levels, the formula is:

$$\frac{R1}{R2} = \frac{V_r + 0.6}{V_{out} - (V_r + 0.6)}$$

which gives the ratio of R1 to R2, though not exact values. Taking the example of a 5V stabiliser of the 7805 type being used this time to provide a 8.5V supply, the requirement is:

$$\frac{R1}{R2} = \frac{5.6}{8.5 - 5.6} = 1.931$$

and the next step is to try some values of R2 and find the corresponding values of R1. If we aim to have about 1mA passing through the resistors, a reasonable amount now that these resistors do not have to pass a current that will 'swamp' the 4.5mA from the common lead of the IC, then we can aim for a total resistance in the region of 8K5. If R2 = 2K7, then R1 = 1.931 × 2.7 = 5K2 which is close to the 5% value of 5K1. For R2 = 3K3, R1 is 6K37, and if we use the 5% value of 2.4K for R2 then R1 is 4K6, very close to 4K7.

For example, if we use 2K4 for R2 and 4K7 for R1, then the voltage output, using the first formula, becomes:

$$\frac{(5 + 0.6) \times (7.1)}{4.7}$$

which gives 8.459V, as close to 8.5V are you are likely to get with any reasonable combination of resistors.

The resistors R1 and R2 have been shown as fixed resistors, but since the current is low, there is no reason why these should not be replaced by a potentiometer or trimmer. This makes the circuit into a voltage variable supply, and for some applications this can be useful, though for such purposes it is usually easier to make use of one of the many variable-voltage stabilisers that are available, see later.

Increased currents

In general, if higher currents are needed than can be supplied using a stabiliser such as the 7805, then the most sensible option is to use a stabiliser IC which is rated for higher currents, or a hybrid stabiliser such as the 78H05 which consists of a power transistor combined with a stabiliser circuit for currents up to 5A.

The alternative is to use the stabiliser to control a power transistor which is rated to pass the required current. The power transistor will have to be mounted on a large heatsink, preferably not the one that is used for the stabiliser, and since the stabiliser supplies only the base of the transistor it should usually be possible to dispense with a heatsink for the stabiliser itself. The type of circuit is illustrated in Figure 4.15, using two PNP transistors. For

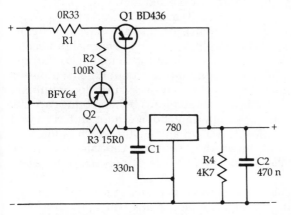

Figure 4.15 A circuit which can use a transistor as a current-dumper under the control of the stabiliser IC in order to provide more current than the IC can pass.

low current values, the IC stabiliser provides the current, since the current through the 15R resistor is not sufficient to provide a 0.6V bias between base and emitter of Q2. The current required is 40mA, so that a small stabiliser can be used such as the 78L types (78L05, 78L12, 78L15) which are intended for a current of 100mA with a dissipation of less than 0.9W. Once the current through R3 exceeds this 40mA limit, Q2 will start to conduct, supplying additional current. When the contribution of Q2 becomes substantial, more than 1.8A, the current flowing through R1 will force Q1 to conduct, supplying most of the base current for Q2 rather than the resistor R3.

In this circuit, the transistors take no part in stabilising the voltage, and are used only as 'current-dumpers', assisting the stabiliser and totally under the control of the stabiliser.

Protection circuitry

In the early days of IC stabilisers, the failure rate of stabiliser chips was very high, but later versions have added protective circuitry which has almost eliminated failures of the type that caused so many problems at one time. The main protective measures are for thermal protection and foldback overload protection. Since these additions to the circuitry are built into the IC and not in any way alterable by the user, it might seem pointless to look at them, but some knowledge of what goes on inside the chip can be useful, particularly to explain why an unexpected failure happened.

The universal stabilising circuit makes use of a comparator amplifier controlling a power transistor, and all protection circuits operate on the comparator amplifier circuitry. Thermal protection makes use of diodes in the bias path of the amplifier so that the bias is reduced as the temperature increases until the chip passes little or no current. This protects well against long-duration overloads or high ambient temperatures which bring up the temperature of the IC fairly uniformly, but it cannot protect so well against transient overloads, and in any circuit which is likely to cause sudden current overloads a reservoir capacitor should be used on the stabilised side in order to supply such transient demands.

Foldback protection is a method of protecting against damage caused by excessive current. Simple circuits which act to protect the stabiliser will limit the current to its maximum value only when the output is short-circuited, Figure 4.16(a). The stabiliser will then be passing its full rated current, and the dissipation of the stabiliser and of its protection circuits will be large. The better alternative is the 'foldback' characteristic of Figure 4.16(b), in which the current is reduced as the voltage drops to zero because of a short circuit. A simplified version of the type of circuitry that can achieve foldback is illustrated in Figure 4.17. The sensing resistor R_s has its value (of the order of 1Ω or less) chosen so that with the output short-circuited and the current at its short-circuit limit, the voltage across R_s will be enough to keep Q2 switched on, allowing for the dividing effect of R1 and R2. When the output voltage is not zero, however, the dividing effect of R1 and R2 is much more effective, so that current is not so limited.

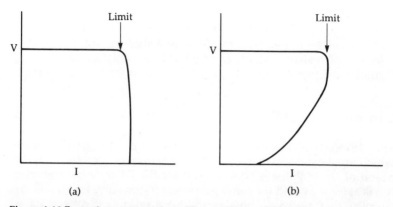

Figure 4.16 Protection systems. A simple protection system behaves as shown in (a), with the current limited, but passing enough current to keep the dissipation high. A foldback system (b) will *reduce* the amount of current to lower than the maximum when a short-circuit occurs.

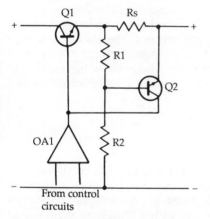

Figure 4.17 A simplified version of the type of circuitry that is used to provide foldback protection.

For example, suppose that R_s is 1R and the division ratio of R1 and R2 is around 0.8. With the output short circuited, a current of 0.75A through R_s will provide 0.75V at the collector of Q1, and a voltage of 0.6 at the base of Q2, switching Q2 on because its emitter voltage is zero, and causing the drive to the base of Q1 to be reduced. With the output level at 5V, the emitter of Q2 will also be at 5V, and to get its base voltage to 5.6V requires the voltage at the collector of Q1 to be 7V (since $7 \times 0.8 = 5.6$). This means that the voltage across R_s can be $7 - 5 = 2V$, and this corresponds to a current of 2A. In this example, then, the critical current is 2A and this will be folded back to 0.75A for a short circuit. By suitable selection of resistor values, the foldback current can be reduced, further reducing the dissipation when a short-circuit fault occurs.

Low dropout stabilisers

The dropout of a stabiliser is the minimum voltage difference which must exist between input and output in order to sustain the action of the stabiliser. The dropout for the 78 series of stabiliser ICs is at least 2V and for some purposes, particularly for stabilising supplies that are based on secondary batteries, this differential is too large. Low dropout stabilisers allow much lower levels of input voltage, typically down to 5.79V for a 5V output stabiliser.

Low-dropout stabilisers were originally developed for the car industry to provide stabilised outputs for microprocessor circuitry, and they would not be used for general-purpose stabilisation for mains-powered supplies. Most low-dropout stabilisers feature additional protection against supply reversal, the effects of using jumper leads between batteries, and large voltage transients. Several types also feature inhibit pins, which allow the stabiliser to be switched back on again after it has been switched off by an overload.

Voltage-variable stabilisers

The fixed voltage stabilisers are used for some 95% of all voltage regulation requirements, but for a few applications adjustable stabilised voltages are required. Most of the adjustable-voltage stabilisers use three-pin connections with a variety of case styles such as TO92, TO202, TO220 and TO3, and one variety uses four pins. The three-pin types use one pin as input and another as output, though in some case styles the output is the connection to the case, and the third pin or third connection is the adjustment. The four-pin style of connection uses separate common (earth) and control pins. Current ranges are much the same as for fixed-voltage regulators, 100mA to 10A, and the same methods can be used to extend current range as are used for fixed-voltage ICs.

A typical circuit, omitting rectifier and reservoir, is illustrated in Figure 4.18. This is very much the same type of circuit as was illustrated for adjusting the voltage output of a fixed-voltage

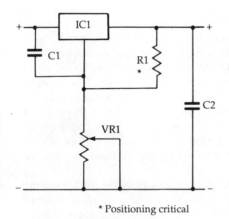

* Positioning critical

Figure 4.18 A circuit that makes use of a variable-voltage stabiliser. The current from the common terminal is much lower for this type of stabiliser than from the fixed-voltage type.

regulator, but the important difference is that the current flowing to earth from the adjustment pin of the stabiliser is very much lower, typically 50µA, and this current is much easier to swamp with the current through R1 and VR1, usually set to be at least 4mA. The formula for output voltage depends on the type of IC that is used, and for the popular LM317 family is:

$$V_{out} = 1.25(1 + \frac{VR1}{R1}) + VR1.I \text{ where I is about } 50\mu A$$

and for this family of stabilisers, R1 can usually be set to 240Ω. As before, if the desired output voltage is known and you need only the value for VR1, then it is simpler to calculate it by another method. Calculate the current I′ flowing through R1 as 1.25/R1 (this will always be the same for any voltage output using the LM317, but remember correct units), and then find the value of VR1 as (Vout - 1.25)/I′. For example, if a 12 V output is needed, then the current through R1 is 1.25/0.24 = 5.2mA. The voltage across VR1 is 12 − 1.25, and with 5.25mA flowing (adding the 50µA from the adjustment connection to the 5.2 mA), then the resistance value is 10.75/5.25, giving about 2.05K. If the output voltage range is to be up to 12V, a 2K potentiometer can be used, and its dissipation will be up to 10.75 × 5.25 = 56.4mW, well within the capabilities of a 0.25W potentiometer.

One very important feature which does not appear in the circuit diagram is the positioning of R1. This must be placed so that it is connected as close to the output terminal of the stabiliser IC as is physically possible. The reason is that any resistance between the output terminal and this resistance is subject to a multiplier effect and will raise the output resistance of the stabiliser. The amount of multiplication is (1 + VR1/R1) which is usually quite high, with typical values of 5 to 15. By contrast, the position of the earthed end of VR1 can be to the earth point of the load in order to improve regulation. The stabilisers that are supplied in the TO-3 casings can be used with separate leads to the casing, one to the load and the other to the resistor R1.

The general features of the LM317 series provide for an output voltage range of 1.25V to 37V, with a drop-out voltage (minimum input-output voltage) of 1.9 to 2.5V. The load regulation is 0.1% and the line regulation is 0.01%, with ripple rejection of 65db. Output resistance is typically 10mΩ and the thermal resistance from junction to case depends on the type, ranging from 160°C/W for the low-current types up to 2.3°C/W for the higher-current varieties.

Additional ripple rejection and short circuit protection can be provided by using the circuit in Figure 4.19. The diodes have to be used because of the effect of the capacitors. If, for example, the output is short circuited, the stored voltage on C1 could cause the stabiliser to try to maintain output, and adding the diode D2 will prevent this by providing a short-circuit path to discharge C1.

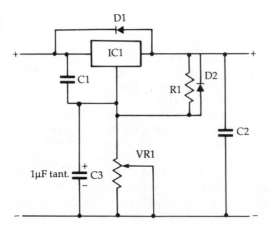

Figure 4.19 Another form of the variable-voltage stabiliser circuit which provides lower ripple and better short-circuit protection.

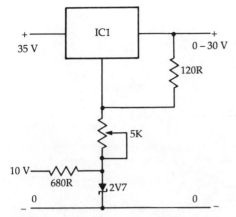

Figure 4.20 A variable-voltage circuit which can be used with either a fixed or a variable voltage stabiliser IC to provide an output whose lower limit is zero.

Similarly, if the input is short-circuited, the capacitor at the output could cause a reverse voltage across the stabiliser which is shorted by D1.

A variable-voltage or fixed-voltage stabiliser can be used to provide a main positive stabilised supply in which the lower voltage limit is zero, but this requires a negative voltage input capable of supplying around 2mA. Figure 4.20 shows the recommended circuit, which uses a 2V7 Zener diode connected to a negative line so as to return the adjustment line of the stabiliser to −2V7. The illustration is of a stabiliser circuit for 0 – 30V, a useful range for experimental power supplies. The transformer/reservoir combination must be capable of supplying at least 35V on full rated current load. Stabiliser ICs are available which can provide for considerably higher voltage ranges (up to 125V) for specialised purposes.

Current regulation

The voltage stabiliser ICs can also be used for current stabilisation, using the circuit of Figure 4.21 which has already been shown in relation to the recharging of nickel-cadmium batteries. The value of the resistor R determines the current, but to this amount must be added the standing current passed through the common connection pin. For example, the 78 series of regulators will pass

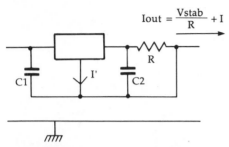

$$\text{Iout} = \frac{\text{Vstab}}{R} + I$$

Figure 4.21 Using a voltage stabiliser circuit to provide current stabilisation in a circuit in which the stabiliser is used 'floating'.

about 4.5mA through the common connection, and for a 12V regulator (7812) the current output of the circuit of Figure 4.20 is 12/R + 4.5mA, with R in units of K. The dissipation in the resistor can be considerable, so that this would normally be a vitreous wire-wound type. For a 7812 0.5A stabiliser, for example, R would be a 24Ω, 6W resistor.

Specialised ICs

There are many more specialised ICs which are of interest to anyone who is constructing or designing power supplies, and of these only a few representative types can be mentioned here. The items that are of particular interest are over-voltage protectors and battery backup switching ICs.

Power supplies that are intended to be used with TTL logic circuitry must guard against over-voltage, which can destroy TTL chips very rapidly. The duration of over-voltage that can destroy TTL chips is much too brief to trigger any conventional fuse, so that only other semiconductor circuits can play any useful part in protecting a circuit against the type of failure of a stabiliser that leads to excessive voltage. As it happens, this is the most common type of stabiliser failure, so that the protection is necessary for any TTL circuit of any significance. Many modern digital circuits make extensive use of MOS devices which are less susceptible to damage from over-voltage, but it is unusual to find a large digital circuit which does not contain at least one or more TTL devices

The typical over-voltage protector chip is a sensor which has to be used in conjunction with a thyristor 'crowbar' circuit. A crowbar, as the name suggests, is a short across the output of the power supply which will force fuses to blow and foldback-current circuit in the path of the excess current (if still active) to come into operation. The principle is that the sensor triggers a thyristor which will in the time of a microsecond or so short the power supply output, reducing the voltage and protecting the circuits. Some few milliseconds later, fuses will blow, completing the protective action and ensuring a complete shut-down of the system.

A protective circuit, due to Motorola and using their OVP (over-voltage protection) chip, is illustrated in Figure 4.22. The minimum

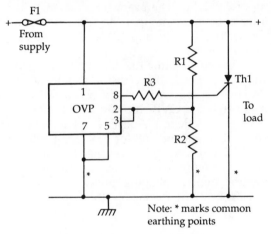

Figure 4.22 An overvoltage protection circuit, courtesy of Motorola, using their OVP chip.

triggering voltage is approximately 2.6V, so that the formula for the trip voltage in this circuit is:

$$V = 2.6(1 + R1/R2)$$

with the value of R2 preferably being less than 10K. Rearranging this gives:

$$\frac{R1}{R2} = \frac{V - 2.6}{2.6}$$

which provides the value for the ratio of resistors. For example, if the circuit is to be used with TTL circuits in which the overvoltage level is to be set to 5.4V, then the ratio is 2.8/2.6 = 1.076, almost equal values. If, for example, we make R2 = 4K7, this would require R1 to be 5K06, and you have the choice of using another 4K7, so lowering the triggering voltage to 5.2V, or to make up the resistance value with a 330R in series. Close-tolerance resistors must be used for R1 and R2, because otherwise the triggering level may be below the normal power supply level, or too high to be useful. The alternative is to use wide-tolerance and build up the values by using series and shunt resistors until exact values, as measured on a bridge, are reached. Since the 2.6V trip level is also approximate, however, the exact triggering level should be determined experimentally by using a variable output supply. The trip voltage level is affected by temperature changes, with a temperature coefficient of 0.08%/°C. The propagation delay of the OVP chip is of the order of 500ns. Note that the circuitry around the OVP chip should be earthed directly at pins 5 and 7 of the chip rather than to any more remote earth.

The resistor R3 controls the current into the gate of the thyristor, and for maximum triggering speed, this resistor should be the minimum possible value that can be used. The value is determined by the amount of current that the OVP chip can supply without damage, so that the value of R3 depends on the voltage level of the normal supply voltage, being about 30R for protecting 20V. At voltages of less than 10V the resistor R3 is not needed, and for voltage levels above 35V, a different circuit is used in any case.

The circuit for protecting higher-voltage supplies is shown in Figure 4.23. This uses a Zener diode to stabilise the supply to the

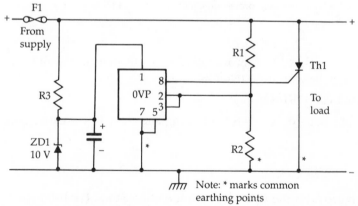

Figure 4.23 Using the OVP chip to protect supplies whose voltage is higher than the normal range of the IC.

OVP chip at +10V, so that no series resistance is needed in the gate connection of the thyristor. The value of R3 for this version of the circuit should be chosen so as to pass 25mA, of which the OVP chip takes a normal steady current of 5mA. The triggering current of up to 300mA is provided by the capacitor which must be connected across the Zener diode.

One problem that affects the operation of any OVP circuit is electrical noise. False triggering because of noise transients can be reduced or eliminated by connecting a capacitor from pins 3 and 4 of the OVP to local earth. The value of this capacitor can be in the range 100pF to 1μF for delay times in the range 1μs to 10ms. A momentary over-voltage will start this capacitor charging and triggering will occur when the voltage reaches the reference level of 2.6V. If the transient voltage reduces before this level is reached, the capacitor is discharged at about ten times its charging rate so as to reset the delay action for the next transient.

Two other features of this chip are indicator action and remote triggering. Pin 6 can be connected by way of a load resistor to supply voltage, and the output from this pin can be used to operate a flip-flop (powered from an unstabilised supply) which can be used to shut down the stabiliser. In this way, a stabiliser which uses no foldback limiting, or in which the foldback current is still high, can be shut down so as to reduce the thyristor current, so making it unnecessary to use a heatsink for the thyristor. This also ensures that the fuse will not blow, and since the unstabilised voltage will still be present, this pin can also be used to operate an LED indicator to show that the OVP action has been triggered. The other provision is that a voltage of 2V or more on pin 5 of the OVP chip will trigger a shut-down whether any over-voltage has been sensed or not. This can be used to ensure a power shut-down when another part of the circuit becomes faulty, so preventing damage that might not otherwise cause over-voltage or over-current conditions.

The comparatively simple type of OVP is likely to be all that is needed for protection of most types of digital circuits, but more elaborate protection chips such as Silicon General's Power Supply Supervisor chip are available. This chip will add foldback current limiting if this is not provided for in the stabiliser, along with sensing of over-voltage and under-voltage conditions, indication of type of fault and crowbar protection when used along with a thyristor.

Another type of action is the automatic switching between a mains supply and a battery-backup. Simple circuits can make use of a diode in series with the battery supply, biased off when the mains supply is in operation, but a considerably more advanced form of action is obtained when a specialised chip such as that from Harris Semiconductor is used. This will switch completely automatically between mains and battery supplies, connecting the circuit to whichever supply offers the higher voltage level. The input voltage level can range between −0.3 V and +18V, and the current capability of 38mA can be extended by using the chip to control power transistors.

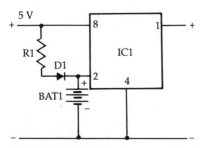

Figure 4.24 The Harris Semiconductor automatic battery backup chip in a typical circuit.

Figure 4.24 shows a typical application in which the backup is carried out by a rechargeable battery, trickle-charged by way of a diode and limiting resistor when the mains supply of 5V is active. On switch-off or failure of the mains supply, the battery supply will be used with the output voltage virtually equal to battery voltage (no 0.6V diode drop). This lower level can then be used for retaining memory, or for whatever purpose is required. A long-life lithium battery can be used in place of a rechargeable battery, in which case no recharging resistor or diode is needed.

5 Switch mode power supplies

The traditional approach to the design and construction of a stabilised supply is in many ways far from ideal. To start with, a large amount of energy is wasted because the unstabilised output level has to be maintained considerably greater than the stabilised level, and the voltage difference will result in considerable dissipation of heat from the stabiliser. At low voltage levels in particular, very large values of capacitance are needed for the reservoir, and this demands electrolytics of 100,000µF to 500,000µF, with all of the problems that attach to electrolytics. The low frequency of the ripple from a mains-operated supply is always difficult to remove completely, even using large capacitors, unless the voltage level is high enough to allow inductors with their inevitable series resistance to be used.

The answer to many of the problems of making high-current low-dissipation low-voltage supplies is the use of switch-mode supplies, which exist in several types. The types that are used for applications such as TV receivers use the mains at full voltage to provide power for an oscillator whose output in turn is rectified and stabilised. Another option, used particularly for digital equipment, is to employ a step-down transformer to provide mains isolation, rectifying the output of this transformer to use as the supply to the switch-mode circuits. When this latter approach is used, the circuit can provide for a step down or a step up of the DC voltage applied to the chips, using an inductor when a step up is needed. These circuits also generally make use of higher frequencies for operation.

When full mains voltage is used as the source of the DC for the switch-mode circuit, the isolation of the supply may have to be carefully considered. For TV receiver use it has been customary in the past to allow a receiver to have its chassis connected directly to the mains. This has fallen out of favour at last because of the need to use TV receivers for monitor actions, and because the direct video output of a VCR can provide much better viewing of video images when direct connection is possible. In addition, most of the voltages that are required for a modern TV receiver are low voltages, and the few supplies that are necessarily of high voltage can be obtained by use of DC-DC conversion techniques (see Chapter 6 also). Though most modern designs feature fully isolated power supplies, there are large numbers of TV receivers still in use which have live chassis working. The form of switch-mode supply for such circuits is the simplest, and an outline of the process is illustrated in Figure 5.1.

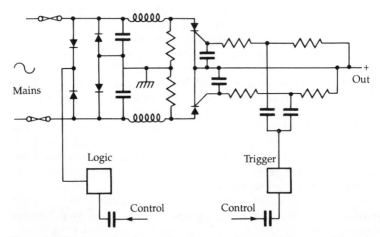

Figure 5.1 The form of switch-mode power supply as used for many TV receivers, with no mains isolation, and with switching at twice main frequency.

In this arrangement the mains voltage is rectified in a bridge circuit, with the negative output of the bridge connected directly to the local chassis earth. This makes it necessary to feed the receiver from an isolating transformer in the course of servicing, and to take great care with the use of instruments which have earthed chassis construction. The positive output from the bridge consists of positive half-cycles at twice mains frequency, and this is used to form the trigger pulses for the switching. Unlike later types of switching circuits, the switching rate is low, so that the need for large smoothing capacitors is not removed, and the main benefit of using the circuit is the elimination of transformers for supplying low voltages. The logic circuitry allows for the regulation of the supply and for foldback of current in the event of a fault condition. The triggering circuits then generate a trigger pulse from the rectified pulses at twice mains frequency and use this to switch the two thyristors. The thyristors operate from the mains input with an output which consists of steep-sided pulses, which can be smoothed and used as a supply at a voltage level of considerably less than the mains supply voltage.

The disadvantage of using low-frequency switching is that smoothing is almost as difficult as it would be for a conventional stabilised supply, but this can be assisted by using *active smoothing*. An active smoothing circuit is similar in outline to that of a series stabiliser, with a reference voltage supplied from a circuit that uses a resistor and capacitor for smoothing. Since this reference voltage needs to supply only a negligible current, the resistor and capacitor can both be of fairly large value, ensuring excellent smoothing of this voltage, and the negative feedback action of the stabiliser circuit ensures that the main supply is also smoothed − you can think of an active smoothing circuit as being a DC amplifier whose input is a perfectly smooth voltage.

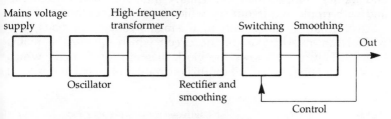

Figure 5.2 A block diagram for a more modern form of switch-mode supply, using a high-frequency oscillator and transformer.

The more modern form of switch-mode power supply is illustrated in the block diagram of Figure 5.2. The mains voltage is rectified, using a bridge set of diodes connected directly across the mains, and this output is partially smoothed. The rough DC is used to operate an inverter oscillator whose output is connected to a high-frequency transformer − a typical frequency is 50kHz so that the transformer can be a comparatively small and light component. All stages up to and including the primary of this transformer are live to mains.

At the secondary of this transformer, low voltage windings can be used, and these will not be live to mains. The output from a winding can be rectified, using a bridge circuit, and smoothed using comparatively low-value capacitors. For low-current lines, a conventional IC stabiliser circuit can then be used, but more usually a switching circuit will be used, chopping the DC into square waves whose mark-space ratio (ratio of high voltage time to low voltage time) can be controlled by a voltage-controlled oscillator. This square waveform is again smoothed into DC, and the level of this DC is used to supply the control voltage for the switching circuit.

In this way, stabilisation is carried out with minimum loss of power, because the switching circuits will be either fully conductive or fully cut off, and changes in the stabilisation conditions do not cause large changes in dissipation. Heatsink requirements are negligible, and the design of the circuit makes it easy to include cut-off provisions in the event of over-voltage, over-current or overheating. In this form, however, the circuit contains some redundancy in the sense that smoothing is being done twice and the square wave formation is being carried out twice, once in the inverter stage and again in the switching stage. The benefits are precise control of low-voltage high-current supplies, with smaller and lighter components (particularly inductors and capacitors), and with ripple at a frequency which is easily smoothed.

An alternative, still preserving the mains-voltage approach, is to control the inverter part of the circuit which operates at mains voltage. Figure 5.3 shows the block diagram for a circuit which uses an opto-isolator circuit for this purpose. The usual mains-level

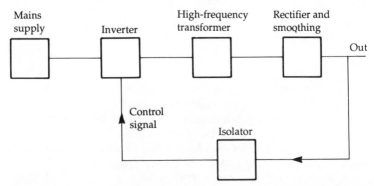

Figure 5.3 Another form of switch-mode circuit in which control is exerted on the oscillator by way of feedback through an opto-electronic link.

bridge rectifier and reservoir circuit operates an inverter whose mark-space ratio can be controlled by a DC signal. The output of the inverter is converted to the correct voltage level using a high-frequency transformer as before, and this is rectified and smoothed. The smoothed output is used to control the inverter by way of an opto-isolator acting to supply the DC control voltage from the inverter using as its input the DC output from the stabilised voltage. In this way, the output side of the circuit is not at mains voltage, but can still be used to exert control on the inverter circuit which is at mains voltage.

This approach is comparatively simple in block form, but because of the losses in the opto-isolator it can require rather more circuitry than might appear to be needed. Another problem is that some regulatory bodies concerned with electrical safety do not consider opto-isolators as suitable for total insulation between mains supply and the chassis of electronic equipment. The problem is that most opto-isolators have only a small separation between input and output, less distance than is stipulated to be used between live parts in electrical safety regulations.

The use of pulse-transformer methods is an alternative that can more easily pass electrical safety approval tests, since a transformer can be manufactured to any required standard of isolation. In addition, the use of a pulse-transformer method obviates the need to have an inverter working at mains voltage, and substitutes in its place a pulse amplifier.

An outline block diagram is shown in Figure 5.4. The power supply that is derived from the mains is used as the supply to a

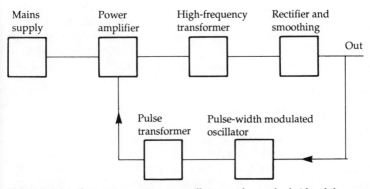

Figure 5.4 Another option, using an oscillator on the earthed side of the system, with a pulse transformer feeding high-frequency signals to an amplifier on the mains side.

pulse amplifier whose input drive is obtained from the secondary of a pulse transformer. The output from the pulse amplifier drives the main high-frequency transformer, whose output is at a low voltage and which is rectified and smoothed in the usual way. This voltage is used as the input to a pulse-width modulated oscillator, and the pulses, whose width will be inversely proportional to the output voltage of the supply, are used as driver pulses to the pulse transformer. Though two transformers are used in this circuit, the driver pulse transformer can be a relatively small component, though it must be as stringently isolated between primary and secondary as the main high-frequency transformer. A usual requirement is a long-duration soak test with 3.5kV or more between windings.

The performance of either type of switch-mode supply can be very impressive. A typical example is a 100W unit from Weir Electronics which offers three fully-regulated and two semi-regulated outputs. This supply can be obtained in three variants with different output ranges, and the figures quoted are for the HSS 100/2 which supplies +5.1V at 12A, +12V at 5A and −12V at 2A, with subsidiary supplies of −5V at 1A and +12V at 2A. The +5.1V 12A supply is protected against over-voltage, and if the mains supply falls to a level that makes stability impossible to achieve, a power failure signal lamp will illuminate. All outputs are current-limit protected against short circuits, and the level for over-voltage protection is 6.2V. Line voltage regulation for a 15% line voltage change is 0.1% to 0.5%, depending on the output that is being measured, and the load stabilisation for a load change of anything from 20% to 100% is from 2.5% change to 0.25% change, depending on the output measured. At a full 100W output, the residual mains output ranges from 5mV to 25mV, the switching-frequency output from 25mV to 60mV, and transient spikes from 50mV to 120mV. The unit can be uprated to 150W if forced-air cooling is available, and there are also other units which are rated at 200W and 300W respectively with natural cooling.

Low-voltage input circuits

The type of switch-mode supply which operates directly from mains voltage with no transformer working at mains frequency is now very common for low-voltage large current use, but it is by no means the only form of switch-mode power supply in use. By making use of a mains-frequency transformer so that the AC input to a power supply circuit is at a low level, switch circuits can be constructed in IC form, considerably lowering cost at the expense of the larger size and weight of the mains-frequency transformer. This approach is, however, very convenient for the designer and constructor of digital circuits which are not in the 5V 10A class and for which the more elaborate supplies which have to be constructed from discrete components are not applicable.

The type of IC that is used is described as a switching-regulator as distinct from a switch-mode supply, and it can be applied to any form of supply, even to battery supplies. Switching regulators can be used to step voltage up as well as down, and so are particularly useful when several different DC supply levels are required from a single DC source, or for use as converters in providing various DC levels from a battery supply, or as inverters for providing a negative supply voltage from a positive supply of the same voltage level.

The 78S40 from National Semiconductor

This is an IC which can be used for step-up or step-down supplies in the range 2.5 to 40V with a current capability of 1.5A. Like conventional IC stabilisers, these switching regulators can be used in conjunction with discrete high-voltage power transistors to extend both the voltage and current capabilities of a switch-mode supply. The power dissipation at zero load is very low, and the efficiency very high, with an internal maximum dissipation of 1.5W.

Figure 5.5 shows the connection diagram for the chip, a 16-pin DIL package. The actions of the various inputs and outputs are

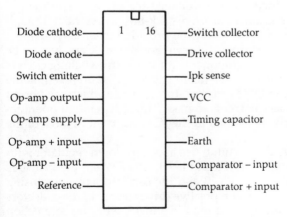

Figure 5.5 Connection diagram for the 78S40 switching regulator IC.

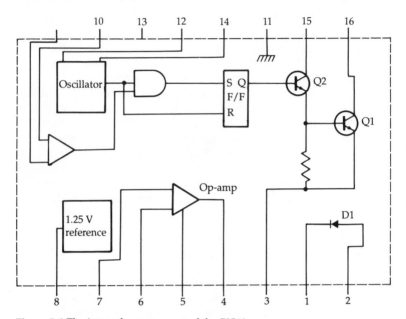

Figure 5.6 The internal arrangement of the 78S40.

illustrated in Figure 5.6, which shows that the chip contains two operational amplifiers, an oscillator, gate and flip-flop circuitry, switching transistors, a 1.25V reference circuit and a diode which can be connected so as to protect the switch transistor against reverse voltage pulses. Since the regulator circuit will be used along with an inductor when voltage step-up or step-down is needed, protection against inductive back-EMF is essential, though for some purposes this can better be provided externally, using a Schottky diode or a silicon fast-recovery diode. Care has to be taken in the selection of components for use with switching circuits of this type, because at the switching speeds that are used, some diodes allow unacceptable levels of stored charge which can permit high reverse transient currents to pass. The independent operational amplifier which is included in the IC is not used in most of the applications.

Figure 5.7 shows the circuit in use as a step-down regulator with a 25V input level and a 5V output level. The 25V supply provides the chip input voltage at pin 13 and the oscillator current sensing input at pin 14, as well as the positive drive supply to the collectors of the switching transistors at pins 15 and 16, with the connection

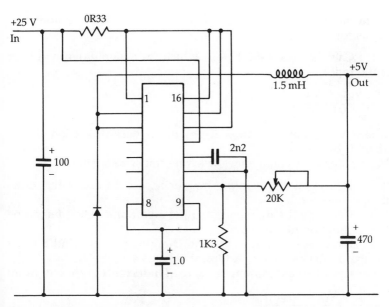

Figure 5.7 A circuit in which the 78S40 is used as a stepdown regulator.

to the diode anode on pin 1 for back-EMF protection. The earth line is connected with pin 11 on the chip, and the timing capacitor for the oscillator is connected between pin 12 and earth. The output is taken from pin 3, the emitter of the final switching stage, through an inductor to generate the voltage pulses for switching and a capacitor to provide filtering. The value of inductance, of the order of a few millihenries, is important, as it determines the ratio of the peak to the average switch current. The larger the value of inductance, the closer these values approach each other, and the value must exceed a minimum that is calculated from the voltage difference between input and output, the peak current, and the on time of each switching cycle. The resistor R1 is of a value that is calculated to limit the peak value of current, and the oscillator is partly controlled by the voltage drop across this resistor.

In use, the DC output voltage across the smoothing capacitor is reduced in the potentiometer-resistor network and compared with the internal voltage reference of 1.25V from a band-gap diode that is part of the IC. The output from the comparator opamp is used to drive a latch that controls the oscillator and its output gate by turning off the drive to the switching transistors when the output voltage is too high.

The steps in the calculation of values for associated components are as follows:

1. Calculate the ratio t_{on}/t_{off}, which for a step-down regulator is the ratio:

$$\frac{V_{out} + V_d}{V_{in} - V_{sat} - V_{out}}$$

where V_d = forward drop across the flyback-protection diode, about 0.7V

V_{sat} = saturation voltage for switching transistors, about 1.1V

2. Decide on an operating frequency, usually above the audio range.

3. Knowing the frequency and the on/off time ratio, determine the on time on microseconds.

4. From the on time value and charging current, find a value for the timing capacitor. A typical value is $4 \times 10^{-5} t_{on}$.

5. Calculate the minimum value of inductance, typically from:

$$\frac{(V_{in} - V_{sat} - V_{out}) T_{on}}{I_{peak}}$$

– and use a value of about double this amount.

6. Use a smoothing capacitor of value given approximately by:

$\dfrac{I_{peak} T}{V_{ripple}}$ for the amount of ripple voltage which is tolerable.

where T is the total time of a cycle.

7. Calculate the voltage divider values to reduce the required output voltage to a level of about 1.25V, adjustable.

This is, of course, only an outline of the steps that are required, making the assumption which is normal for this chip that the peak current will be about twice the average current. For detailed design of a supply using this chip the manufacturer's leaflets, or the data sheets would need to be consulted.

When step-up action is required, the circuit alters to that shown in Figure 5.8. The inductor is now positioned between the collectors of the switching transistors so that its reverse EMF is being used to provide the voltage step up each time the final switching transistor switches off. A separate diode is used for rectification (the built-in diode must not be used for this purpose because of leakage to the substrate of the IC), and the transistors are being used to switch the current in the inductor on and off, rather like the old-fashioned car ignition circuit. In other respects,

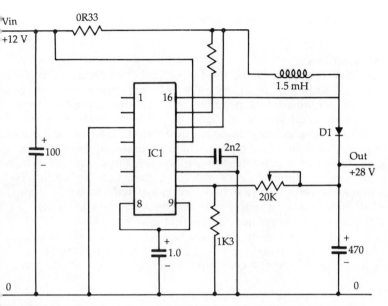

Figure 5.8 The alterations to the circuit so that step-up action can be obtained.

however, the circuit is similar, with the output voltage being divided by the resistors to be compared with the internal 1.25V reference, and the operational amplifier controlling the oscillator. The steps in calculating the value of external components are now:

1. Find the t_{on}/t_{off} ratio from $\dfrac{V_{out} + V_d - V_{in}}{V_{in} - V_{sat}}$

2. Choose an operating frequency, usually higher than the audio range.

3. From (1) and (2), find the value of t_{on}.

4. Find a value for the timing capacitor by using $4 \times 10^{-6} \times t_{on}$.

5. Calculate I_{peak} from $2 \times I_{out} \times \dfrac{t_{on} + t_{off}}{t_{off}}$

6. Calculate minimum inductance value as $\dfrac{(V_{in} - V_{sat}) \times t_{on}}{I_{peak}}$

7. Calculate R1 from $\dfrac{0.33}{1.2 \times I_{peak}}$

109

8. Calculate smoothing capacitance from $\dfrac{I_{out} \times t_{on}}{V_{ripple}}$

Other actions

The 78S40 regulator can be used in several other modes which we shall look at in less detail. As an inverting regulator, the chip can provide a high-current negative output from a positive input, using the circuit shown in Figure 5.9. This requires an external switching transistor and diode because the substrate of the IC must be at earth voltage. One input of the comparator op amp is at earth level, the other is connected to the divided output voltage, which is divided to the reference voltage rather than to earth. The component values shown here are calculated for a 12V input, $-15V$ 500mA output with ripple less than 1% peak to peak.

The switching regulator circuits that have been examined so far make use of controlling the oscillator so as to inhibit switching while the output voltage is too high. An alternative is to keep the oscillator switching at a constant frequency and to vary the mark-

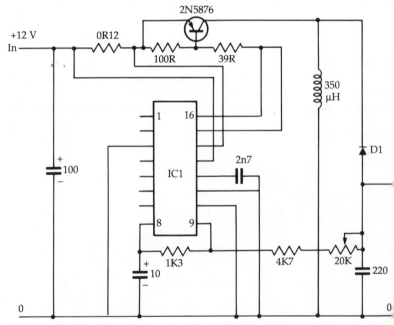

Figure 5.9 Using the 78S40 as an inverting regulator, producing a negative supply from a positive supply.

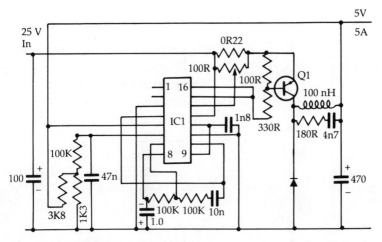

Figure 5.10 A circuit which imposes mark-space control, using the additional operational amplifier.

space ratio in order to maintain a constant voltage output. This requires the use of the additional op amp, in a circuit such as that of Figure 5.10. The timing voltage across the capacitor at pin 12, which is a ramp voltage, is applied as one input to the comparator, with the other input derived from the spare op amp, using as inputs the reference voltage and the attenuated output voltage. The ramp voltage from the timing capacitor will in each cycle reach the voltage at the other input of the comparator, depending on the output voltage level, and this in turn controls the oscillator on/off time ratio.

The L296 SGS-Thomson switching regulator

As an indication of the actions that are incorporated into a dedicated step-down switching regulator, the L296 is a 4A device whose output voltage can be in the range of 5.1V to 40V. The package also includes a crowbar output for operating a thyristor, and programmable current limiting, remote cut-off control, thermal protection, a reset output for connecting to the reset line of a microprocessor and several other features of more specialised interest. Switching frequencies up to 200kHz can be used so that external filtering inductors and capacitors can be very small. The mode of operation is pulse-width modulation, and the IC is contained in an unusual package, Figure 5.11, which incorporates heat-sink mounting.

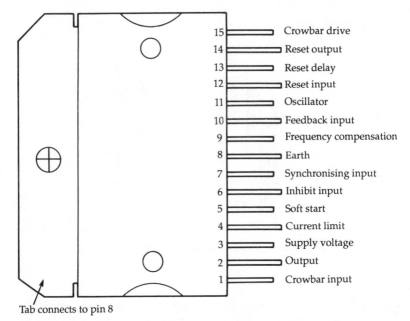

Tab connects to pin 8

Figure 5.11 The physical layout of the L296 switching regulator IC.

The block diagram of this chip is illustrated in Figure 5.12. Pins 1 and 15 handle the crowbar drive, using pin 1 to sense the voltage (usually at pin 10) so that the crowbar action can be triggered when the output voltage is 20% higher than the set value. The output to the gate of the crowbar thyristor is taken from pin 15. The main supply positive voltage, up to 46V maximum, is applied to pin 3, and the earth connection is to pin 8. The output voltage, 5.1V to 40V is obtained from pin 2, with a diode connected to prevent negative excursions when current is switched through the inductor which is used to generate the output pulses.

Pin 4 allows current limitation to be set by way of a variable resistor to earth. If this pin is left open-circuit, the current limit will be 8A, and with 33K in circuit here, the limit is 2/5A. Pin 5 allows the connection of a 'soft-start' capacitor which allows a time delay between the application of input voltage and the appearance of an output voltage. The soft-start capacitor also determines the average output current for short-circuit conditions.

Pin 6 is an inhibit input, so that the output of the regulator is switched off when this pin is taken to TTL high voltage. the input

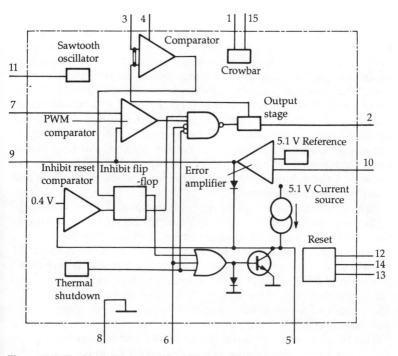

Figure 5.12 The block diagram of the L296 switching regulator.

at pin 7 is for synchronisation when several L296 chips are used in the same circuitry. When this feature is used, only one of the chips should use an oscillator timing network, and all of the pin 7 inputs should be connected together.

Frequency compensation is provided by pin 9, and a series RC network will be connected to this pin to maintain stability in the internal negative feedback loop. Pin 10 is the feedback input to which the divided output DC voltage is applied, leading to the comparator whose other input is 5.1V from an internal reference source. The frequency of oscillation is determined by the RC network connected to pin 11 and also to pin 7. If several chips are to be synchronised, only one will have this connection, and the pin 7 inputs will be connected, with no connections to pin 11 of the other chips.

The reset input is at pin 12, and is of particular interest when the chip is used for the power supply to microprocessor equipment or any logic circuit that can make use of a reset signal. When the

113

voltage on the reset input, pin 12, is lower than 5V the output reset pin is at low level. When the voltage on pin 12 is more than 5V, there is a time delay and then the voltage on the reset output goes high. The reset input pin is normally connected to the feedback input pin, pin 10 at which the normal level is about 5.1V. The reset input can also be connected through a potential divider to the input unregulated voltage so that it can be used as a power failure indicator. The time between the reset output going low and the actual drop of power output can be enough to allow for data to be saved into CMOS memory that is battery-backed. Pin 13 provides for a capacitor to be connected to determine the amount of delay between establishing a high enough voltage at the reset input and switching the reset output.

Figure 5.13 shows a typical application circuit for the L296 chip, with components selected for a 12V output and for each feature of the chip to be used. The voltage output level is set by the value of R8, using 6K2 in this example, with R7 a 4K7 resistor. Resistor R1 is used to set the minimum input voltage for a power failure reset. The ratio of R1/R2 should be equal to $(V_{in}/5)-1$, so that if a 10V level has to be detected, then this gives a value of 1, making R1=R2. The values of the resistors used in this part of the circuit should not exceed 200K

The frequency of operation is set by R3 and C3, and the value of C3 has to be in the range 1nF to 3n3, with R3 in the range 1K0 to 100K. In this example, typical values of 2n2 and 4K3 are used, providing a frequency of about 100 kHz. The resistor R4 is a pull-down resistor for the inhibit pin. This pin can be directly earthed if inhibit action is not needed, otherwise the pulldown resistor should have a value not exceeding 22K.

The network composed of R5, C5 and C6 carries out frequency compensation at pin 9; the parallel capacitor C6 is not needed if the output voltage is 5V. The minimum value of R5 is 10K and in this example 15K has been used. This resistor in conjunction with C5 and C6 will set the high-frequency loop gain, and since the calculation of values is not easy, the recommended values should be used.

The resistor R6 is a load for the reset output and will not be needed unless this output is being used. The value should not be too low, not less than the value which will pass 50mA at normal output voltage, and preferably higher if this does not cause excessive delay in resetting. 1K0 is a reasonable value to try if this action is required.

The output voltage is set by the dividing action of R7 and R8, and the value of R8 must not exceed 10K, so that this component can be a variable of value up to 10K if variability is needed. In the

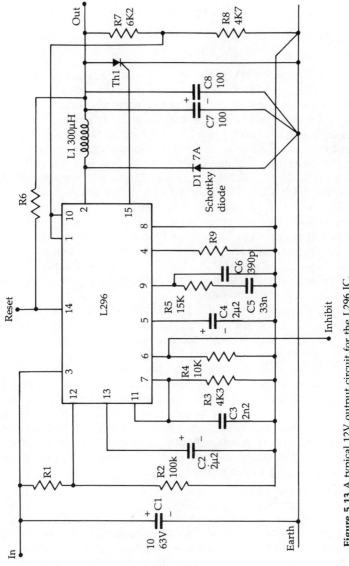

Figure 5.13 A typical 12V output circuit for the L296 IC.

115

diagram, R8 has been fixed at 4K7, with R7 6K2 for a 12V output. The calculated value for the ratio R7/R8 is obtained from:

$$\frac{R7}{R8} = \frac{V_0 - 5.1}{5.1}$$

where the figure of 5.1V is the internal voltage reference level.

The limiting resistor R9 is connected from pin 4 to earth. If this resistor is not connected, the current is internally limited to 8A. A value of R9 of 33K will limit the current to 2.5A, and values in the range 10K to 100K can be used for limiting in the usual range of current outputs. The value should be found by experiment.

Of the capacitors that have not been associated with resistors, C1 is the stability capacitor at the input, and will normally be an electrolytic with a low ESR rating and with capacitance of at least $1\mu F$. The typical value illustrated here is $10\mu F$. Capacitor C2, also an electrolytic, sets the reset delay, and will be used only if the reset action is required. C4 is the soft start capacitor whose value must be at least $1\mu F$. This capacitor value will also determine the size of the average current on short circuit, so that values of less than $1\mu F$ are undesirable from this point of view rather than for making the soft start too rapid.

The final components in the circuit are the output filter capacitors C7 and C8 along with L1. The two electrolytics must have low ESR and self-inductance values, and splitting the capacitance between two capacitors in this way makes it easier to achieve a low ESR value because the series resistances of the capacitors are in parallel. The self inductance values are also in parallel, which also lowers the effective value. The inductor must be wound on a core of very magnetically soft material which will not saturate at currents up to 0.2 times the normal current limit. The value of inductance is set from the input voltage V_i, output voltage V_o, frequency f and allowable current ripple in the inductor, I_f, using the formula:

$$L = \frac{(V_i - V_o)V_o}{V_i f I_f}$$

The typical value obtained in this illustration is $300\mu H$, an inductance value that calls for about 27 turns of wire on the type of core that is likely to be used. For a practical application, of course, the inductance and the number of turns would have to calculated more closely.

The value of capacitance is set after the coil inductance has been calculated with reference to the frequency of operation, input voltage V_i and output voltage V_o, inductance value L and output ripple voltage to be achieved V_r. The formula is:

$$C = \frac{(V_i - V_o)V_o}{8L\,f^2V_r}$$

In this example, the formula has led to the value of 200μF and this has been split into two 100μF units so as to keep ESR to a minimum.

From all of this, it can be seen that using a switch-mode regulator is by no means a simple option, but it is very much simpler than the use of a discrete-component circuit. Considerable help in the form of suggested circuits is available from the manufacturers, including advice on heatsink dimensions and attachment.

As a further illustration of the use of this type of chip, Figure 5.14 shows a variable-output power supply which can be fed from a 20V secondary transformer rectified with a bridge circuit and using a reservoir of about 6000 to 10,000μF. The circuit uses a 5K0 variable in the output-voltage determining stage, allowing variation of output voltage from 5.1V (the minimum level determined by the internal reference voltage) to 15V, the maximum feasible with the transformer voltage specified. The output current is a

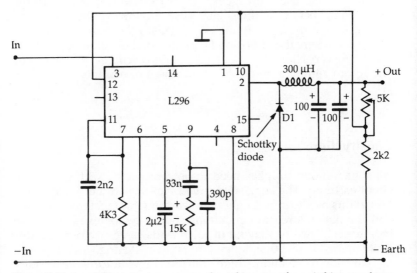

Figure 5.14 A variable-output power supply making use of a switching regulator.

maximum of 4A, but no current limiting resistor has been used so that current limitation takes place at the default value of 8A. There is a minimum load current of 100mA for correct operation, ripple is less than 20mV, and load regulation for output currents of between 1A and 4A is of the order of 10mV when the output voltage level is 5.1V. The line regulation is 15mV for a 15% variation in the input voltage when the load voltage and current are 5.1V and 3A respectively.

Circuits of this type are limited to a lower level of 5.1V because of the internal reference level, and adjustment down to zero voltage output can be obtained only by modifying the circuit which is used for voltage comparison, returning the voltage divider resistors to a positive supply rather than to earth. Another variant on the basic circuit is to use a transformer in place of the single winding inductor at the output. The secondary of this transformer can then be used to provide another output which can be rectified using a Schottky diode, then smoothed to provide a low-current supply whose stabilisation is not of the same order as that of the main supply.

Since each chip can in this way provide both a well-regulated supply and a secondary supply of lesser regulation, two chips can be used in a synchronised circuit to provide the typical computer supply lines of +5.1V 4A, +12V 2.5A, −5.1V 200mA and −12V 200mA. When multiple chips are used in this way, the crowbar outputs can be used to trigger a thyristor which is placed in the input line, between the main fuse (following the reservoir capacitor) and the input to the chips. This will ensure that all lines go down when the thyristor is triggered, but allowing a delay (while the capacitors in the output circuits discharge) which can be enough to allow for an orderly shut-down of memory.

Construction

All switch-mode supplies need to be constructed with care taken about earthing. The suggested circuits for switch-mode regulators hint at this by showing single-point earthing for most of the earths, and the maker's recommendations should be followed closely in this respect. Incorrect layout of earth leads can cause considerable problems in circuits that depend on rapid switching, and can result in parasitic oscillation, high output impedance levels, false triggering of the crowbar circuits and other problems.

RF interference

The use of switching-mode supplies and regulators has considerable advantages in weight and size of the main components and
in particular the smoothing requirements. One disadvantage,
however, is RF interference. The essence of a switch-mode supply
is that large charging and discharging currents are likely to flow,
and where these currents flow in stray capacitances and induct-
ances there is likely to be resonance that can result in RF being
generated. The problem is tackled firstly in the layout of the circuit
by ensuring that earthing, particularly single-point earthing, is
correctly carried out, and circuits are encased in metal screens to
reduce radiation as far as is practicable, but from then on, filtering
is required to reduce the level of conducted RF interference (RFI)
to a minimum.

Filtering is made considerably easier if the circuit uses a mains-
frequency transformer, since this prevents the spread of RFI
through the mains lines. The use of a mains transformer with an
earthed screen between primary and secondary, along with a 10nF
capacitor across the output terminals of the transformer, results in
a reduction of conducted RFI to well below the limits imposed by
regulations of bodies such as the FCC in the USA. Further
reductions can be achieved by using an inductive filter on the
secondary side of the mains transformer, and by filtering each
output of the power supply. The usual range for testing for RFI is
10kHz to 30MHz, but good filtering will ensure that low levels of
RFI are found for frequencies well above the 30MHz limit.

6 Other supplies

DC-DC converters

The conversion of one DC level to another is employed more and more in digital circuitry. Where a step-down of a supplied DC level is required, of course, this can be achieved in a simple way by using a resistor and Zener diode, or with a more elaborate series stabiliser circuit. This, however, involves the waste of energy through dissipation in a resistor, and cannot be applied when the available input DC level is low and the required output voltage level is higher. Both problems can be solved satisfactorily by DC-DC conversion.

The block diagram for such a converter is shown in Figure 6.1, and it closely follows the general layout of a switching regulator. The input DC is used to operate an oscillator whose switching is internally regulated by the control circuits. The resulting square-wave output is converted in level by using an inductor, and the

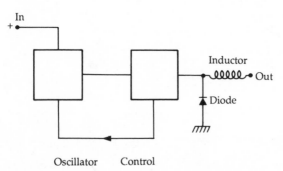

Figure 6.1 A block diagram for a DC - DC converter.

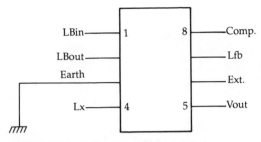

Figure 6.2 The Maxim MAX631 DC - DC converter chip.

waveform is rectified and smoothed. The main difference between an IC intended for DC-DC conversion and a switching regulator is that the dedicated DC-DC converter module is used for low-current outputs, not the outputs of 1 − 4A that we were considering in Chapter 5.

Figure 6.2 shows the pin-out of the MAX631 chip from Maxim. Though the chip uses eight pins, the only external components that are necessary are the inductor and output filter capacitor. With an input of 2V to 16.5V, the output is +5V 135mA with load regulation of 0.2% and input voltage regulation of 0.08%. The oscillator frequency is fixed at 45kHz, and the efficiency of conversion is typically 80%. The minimum voltage for starting oscillation is 1.3V. The chip also contains low-battery indication circuits which allow for low-battery signals to be generated or transferred (so that several chips can have their low-battery circuits chained to an indicator). A charge-pump circuit, see later, is also included so that a negative voltage can be generated by adding external circuitry.

Several chips of this family are available, including negative voltage outputs and 10W versions, all of which follow much the same operating principles and require external inductor and capacitor components. The negative voltage versions also require an external diode, because the built-in diode is connected to the substrate and cannot be used to provide a negative voltage output.

Diode-pump action

One problem of DC-DC conversion is the need for an inductor to perform the voltage step up, but it is possible for very low-current

121

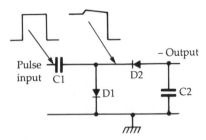

Figure 6.3 Principles of the diode-pump circuit which is used in A-D converters and also as a way of changing signal levels in DC-DC converters,

supplies to achieve a voltage step up without the presence of an inductor, using the diode-pump type of circuit. The principles, which are very similar to those of voltage-multiplier circuits, are illustrated in Figure 6.3 in which the input is a square-wave or pulse waveform.

Referring to Figure 6.3, the capacitor C1 will ensure that the DC level at the input has no effect on DC levels beyond this point. If the diodes were disconnected, the waveform at the output side of the capacitor would be balanced around earth, with both positive and negative peaks. The effect of the diode D1 is to pass current on positive peaks, charging C1 so that the average DC level is negative, and the waveform at the cathode of D2 is a set of negative-going peaks. D2 in turn will conduct on these negative-going peaks, charging C2 to a negative voltage whose amplitude will be the peak-to-peak of the original wave, less the diode drops.

The use of diodes in this simplified version of the circuit makes it impossible for the output to match the voltage of the pulse input, but if MOS devices are used as rectifiers in place of diodes, the drop can be reduced to an insignificant amount. Combining this circuit with a square-wave oscillator in a single chip will make it possible, therefore, to generate a negative voltage supply from an equal positive supply. The current capabilities are limited by the amount of charge that the capacitors can store in the time between charging periods, so that this form of voltage level change is not so useful as a transformer for larger loads, but has the merit that it needs no inductive components. This allows the construction of inverter chips which are used to provide a negative supply for applications in which it would be wasteful to make provision for such a supply on the main power-supply board.

Siliconix Si7661 DC-DC inverter

The Si7661 is a DC positive to DC negative supply converter which is arranged to generate a precisely complementary or inverted supply, meaning that if it is provided with +12V DC it will give a −12V DC output, and if provided with a +5V DC input it will generate a −5V DC output. The only external components that are required in the simplest applications are two capacitors. The internal circuitry of the chip contains a voltage stabiliser, oscillator, level shifting circuit (diode pump), MOS switches, and a logic circuit which ensures that the substrate connections to the MOS switching transistors are never forward biased for any combination of inputs and outputs.

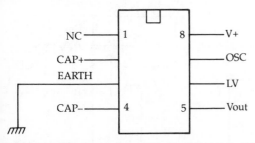

Figure 6.4 The pinout of the Siliconix Si7661 DC - DC inverter, which provides a negative supply of the same amplitude as the positive supply.

The pin-out is shown in Figure 6.4. The LV pin is left unconnected unless the input and output voltages are low (less than 8V). For low-voltage use, connecting pin 6 to earth makes the earth level the new voltage reference level in place of the internal reference voltage. This improves the action for input/output levels of less than 8V, but for higher levels pin 6 should be left unconnected. The internal oscillator operates at a nominal 10kHz (unloaded), but the frequency can be lowered by connecting a capacitor from the OSC terminal, pin 7, to earth, or the internal oscillator can be ignored and the circuit driven from an external oscillator through this OSC pin.

The simplest basic inverter circuit is illustrated in Figure 6.5. The supply voltage is applied to pin 8, with pin 3 earthed, and pins 2 and 4 are used for the pump capacitor, using an electrolytic capacitor of a value which must be less than 100μF − the typical

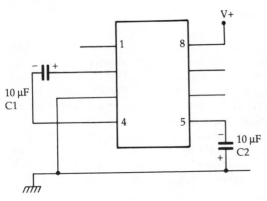

Figure 6.5 The simplest possible inverter circuit using the Si7661.

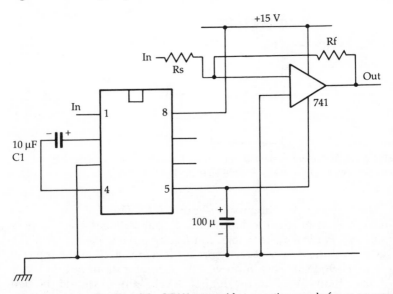

Figure 6.6 An application of the Si7661 to provide a negative supply for an op-amp used on a board with only a positive supply provided.

circuit here shows a 10μF capacitor. The inverted output is from pin 5, and this is smoothed by another externally connected capacitor of the same size, taking care to observe correct polarity.

Inverters of this type are not intended to provide substantial load currents, nor are the outputs stabilised to any extent. They are of particular use when a device needs a negative bias line that uses little current and does not warrant the expense of an additional power supply winding or the use of an additional Zener diode stage. The usefulness of the inverter is that the chip can be placed on the same board as the IC that requires the negative line, so that the board as a whole operates from a single voltage line. This is particularly valuable if the board is intended to plug into equipment that does not supply negative lines and cannot be modified to do so.

As an example of how simple this application of the Si7661 can be, Figure 6.6 shows the chip being used to generate a negative supply for a 741 op-amp in a board that uses only a positive supply line. The 741 should be operated with low current outputs because a large current drain will cause the output voltage of the inverter to drop towards zero. Inverters can be paralleled to provide more current, and problems of beating of oscillator frequencies can be solved by synchronising the oscillators, but when this type of measure is contemplated it is usually better to make other arrangements for the negative supply.

EHT supplies

There is no precise definition of what is now meant by EHT (extra high tension), but the original idea was that valve-operated equipment could normally be expected to use voltage levels up to 500V DC and anything higher than this was EHT, whether for a valve transmitter circuit, a TV cathode-ray tube, X-ray tube, or high voltage test equipment. Some of these EHT supplies such as those used for radio transmitters or X-ray equipment demand very substantial currents, others use less than 1mA, so that the design of EHT power supplies varies considerably and some varieties are so specialised as to be beyond the scope of this book.

As an example, large radio transmitters may call for a 20kV supply at several amperes of current. This order of power is well outside the scope of semiconductors, and calls for three-phase transformers with either hard-vacuum diodes or mercury-vapour rectifiers. The use of three-phase eases the smoothing problem, which uses large inductors and oil-filled capacitors whose size can approach that of a small room for even a quite modest capacitance. X-ray equipment may require 100kV at a current of less than 1A, delivered usually by a half-wave rectifier that consists of a hard-vacuum (as distinct from mercury vapour or other gas filling) valve using a platinum anode and a tungsten or platinum filament. Smoothing at this level is minimal and of little importance in any case.

As far as the general run of small-scale electronics is concerned, EHT supplies are used mainly for cathode ray tube (CRT) anodes and for some specialised purposes such as Geiger-Muller counters and photomultipliers. None of these applications calls for a large current drain, and though in the distant past it was quite common to make use of mains transformers and half-wave rectifiers (valve or solid-state), the more usual methods now are the use of inverters along with semiconductor diodes. The methods that are mainly used have been derived from the standard form of TV circuitry employed for many years.

The circuit of Figure 6.7 shows a portion of a line-output stage for a computer monochrome monitor which uses a BU806 line driver transistor, which has a built-in diode to prevent the collector becoming negative with respect to the emitter. The collector is fed from the 12.5V positive line through a diode and the primary of

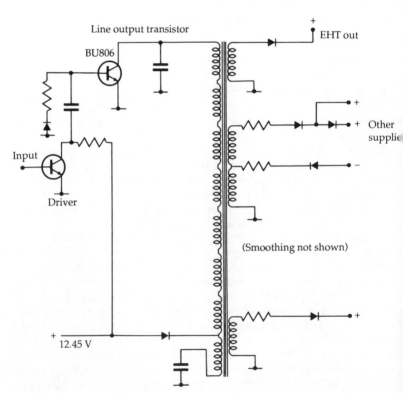

Figure 6.7 A typical EHT circuit for a small monochrome monitor, using a single winding of the line transformer and a single diode for rectification.

the line output transformer, with the connection of the 12.5V supply taken to a tap on the transformer. When the transistor is pulsed to cut-off, the effect of the transformer's inductance is to generate a large positive-going pulse at the collector of the transistor. This is used as the output to the deflection coils, and several secondary windings also make use of this pulse to generate voltages ranging from 60V through 500V to the 12kV needed for the CRT. This EHT winding is better insulated than the others, and is rectified by a diode of specialised construction which should be mounted in an insulated holder. Smoothing of the output, which has a ripple frequency of around 15kHz, is carried out by the stray capacitance around the anode of the CRT.

The EHT power supplies for colour receivers must be able to supply higher levels, of the order of 16kV to 24kV, and these circuits, though they are essentially similar in outline, make use of voltage multiplier techniques. The most common method is to split the EHT windings into several sections, with diode rectifiers placed between the windings as were previously illustrated in Figure 3.16. In addition to higher voltage requirements, colour receivers need more current and also some degree of stabilisation. For TV use, the stabilisation has to be provided at minimal cost, and the most common provision is a shunt stabiliser using a non-linear resistor of several megohms nominal value. The current requirement is met by using larger currents in the primary of the output transformer so that the amount of power that is switched is substantially higher.

EHT supplies for laboratory use require considerably better stabilisation than those used for TV receivers, but the general principles follow much the same lines. An EHT inverter consists of a pulse oscillator driving an output stage and making use of the large back-EMF that is generated when current is shut off. This output is rectified, using multipliers if necessary, and a fraction of the output voltage is fed to a comparator and used to stabilise the output by controlling the oscillator. The circuitry is discrete but it follows the same principles as are used in switching regulators.

Any work on EHT supplies should be carried out with very great caution, because capacitors in the circuit may have been charged to several kilovolts. Even though EHT supplies may be current limited and capacitor values small compared to the electrolytics that are used in low-voltage supplies, the discharge of a capacitor can represent a large amount of energy which can prove fatal. Remember that the energy stored in a capacitor is given by $0.5CV^2$ and the squaring of the voltage value results in very large stored energy values in EHT circuits. Switching off and discharging capacitors may not be safe either, because some types of capacitors exhibit a form of voltage hysteresis, so that after being discharged they can build up voltage again, and when work is to be carried out on an EHT supply, all capacitors should be discharged completely. The usual method of doing this is by way of an insulated screwdriver, with the tip held against an earthed point and the shaft touched against the capacitor live terminal. If capacitor discharging has to be done frequently it is better to make discharging tongs, consisting of a polystyrene rod with a piece of stout wire glued to the end. Discharging should be done after the supply has been switched off, and then repeated one minute later.

Exotic supplies

The heading of exotic supplies covers types that are not likely to be encountered by many readers, but which are nevertheless extremely important and might in the future be more widely used. The main subdivisions are fuel cells, nuclear cells, photo cells and thermal cells, all of which are already used quite extensively for specialised purposes. All of these cells have rather low conversion efficiency, and provide low-voltage outputs. Their applications are in general restricted to uses in positions which make the use of other forms of energy conversion unsuitable. It's not a simple task, for example, to change batteries on a satellite or to take mains supplies to a sub-sonic transmitter several miles under the surface of the sea, and it is in these types of hostile or inaccessible positions that most use is made of the types of supply described here. The first example, however, is very much a terrestrial form of supply.

The sodium-sulphur cell was one that was regarded as a hot favourite in more ways than one for powering electric vehicles, though interest has subsided to some extent. The cell depends on the achievement of high temperatures before it can be useful, so that it is quite unlike any other primary or secondary cell in requiring to be heated so that the chemicals are molten before it can be used. The cathode is molten sodium which, like lithium, must be kept out of contact with air or moisture. The sodium is contained in a porous aluminium-oxide ceramic pot which is surrounded by molten sodium sulphide and in turn by the anode, a mixture of molten sulphur and carbon grains. The temperature required for operation is in the range 300°C to 350°C.

At this temperature the aluminium-oxide ceramic material is porous to the positively charged sodium ions, and the sodium sulphide is also a conductor in its molten state, so that the reaction of the cell is the combination of sodium and sulphur to form sodium sulphide, the electrolyte. This is a reversible reaction under the conditions in which the cell is used, so that the cell is rechargeable. The EMF is around 2.1V, comparable with a lead-acid cell. The requirements for sealing against air and moisture along with the high operating temperature make this cell too difficult to use for its original intended tasks of powering electric vehicles. Sulphur-zinc cells of similar construction have also been developed, but the operating temperature problem still remains even though the materials are slightly less of a hazard.

The fuel cell makes use of a chemical reaction that is at first sight very different from the dissolving of a metal by an acid or alkali. Chemically, however, any oxidation of a material involves the release of electrons, even if this release is only an intermediate stage in a reaction and is never obvious in any electrical sense. Burning a fuel is an oxidation process, but the normal form of combustion that involves high temperatures and flame cannot be made to yield electrons to a conducting circuit, because burning involves gases and non-conducting liquids. Hot gases are conductive if they contain ions, but the conductivity is not generally enough for use in a cell. Burning, however, is only one form of oxidation, and cold-oxidation processes, of which one familiar example is the rusting of iron, can be used in way that make cell action more feasible.

In particular, when a combustible material meets oxygen at a surface of a suitable metal catalyst, oxidation can take place at a low (25°C) temperature, with the release of electrons from the metal. Only a few metals can perform this task of catalysis, in which the metal is chemically uninvolved and acts only to provide a place for the reaction to happen. The most suitable catalyst metals are the platinum/palladium set of metals, which are familiar to chemists as catalysts for a host of other reactions, including the catalytic conversion of car exhaust gases to the completely oxidised form. All of these metals can be obtained in spongy form with a huge surface area at which the catalytic action can take place.

Early fuel cells used hydrogen and oxygen gases and were able to provide a small EMF at a rather high internal resistance for limited periods. The drawbacks were that the gases had to be supplied at high pressure, liquid electrolytes were also needed, efficiency was very low, and storage of gases made the whole device cumbersome. Since then, much effort has gone into making fuel cells that can make use of liquid fuels, but purity of liquids is always a problem, and such cells do not compete with dry primary cells for any normal electronics task. The hydrogen-oxygen gas system has applications as a power source for spacecraft, however, and several satellites have made use of this source. Natural-gas rigs can also use fuel cells to power electronics communications and control equipment. The main hazard, once a fuel cell has started running successfully, is poisoning of the catalyst by traces of materials such as lead.

Nuclear cells make use of the natural radiation of electrons by many radioactive materials — they can reasonably be described as the most natural source of electricity known to us. The emission of negatively charged particles from natural radioactive rocks was, in fact, known before the existence of the electron was verified, so that this effect is often called by its older name of beta-radiation rather than as electron emission. Natural radiations were originally identified as alpha, beta and gamma rays. Of these, alpha radiation was found early in the century to consist of positive ions which are fragments of the helium atom. These fragments are comparatively large and would cause harm to living tissue, but because of their size are stopped by almost any material, including air and paper. They travel appreciable distances only in a vacuum. Beta particles are electrons and because of their small size cause much less damage to living materials. Their range is a few mm in air, less in solid materials. Gamma rays are the most dangerous form of naturally-occurring radiation with a longer range of several centimetres in air and severe effects on living tissue. They can be blocked by lead or other dense materials and like all waves, obey an inverse-square intensity law, meaning that doubling the distance between yourself and the source will reduce the received energy to one quarter of its previous value. Waves of equal intensity and danger, called cosmic rays, are also bombarding us continually from outer space, because the whole Universe is radioactive — nuclear-free zones are a figment of the imagination.

The electron emission has a very short range in air, a few millimetres, so that making use of this source of electrons requires the material to be in a vacuum, or arranged so that the electrons can be collected as they leave the material. If, for example, the radiating material is in the form of an insulating ceramic, the collector can be a sheet of metal on the surface of the ceramic. The rate of flow is very small, a matter of picoamperes rather than microamps in many cases, unless the material is very strongly active. To put this into perspective, most samples of low-level radioactive materials have such a low emission rate that the electrons can be counted individually. On a scale of personal danger from 1 to 100, any low-level radioactive waste comes at about 2, a farmer's dung-heap at about 15.

Nuclear cells are used in spacecraft and satellites, and in some undersea and remote light-buoy applications. Their advantages are a very long active life, at least as long as the half-life of the material. The rate of emission for a radioactive material declines exponentially like the discharge of a capacitor through a resistor, so that a form of time-constant, the half-life, is used as a measurement. The half-life is the time that the material takes to decay from one level of activity to half of that level, so that in a time of two half-lives the activity will have dropped to one quarter of the original value, in three half-lives to one eighth and so on. The usefulness of a nuclear cell after one half life depends on the current demand, and a cell will be for all practical purposes exhausted when it can no longer supply enough current for its applications.

The word *photocell*, in the sense of generators as distinct from detectors, refers nowadays to the silicon photovoltaic cell, which use the energy of light to separate electrons and holes. The important point is that light, as distinct from bright sunlight, is required so that diffused light or reflected light can be used and such cells can still provide an output even on cloudy days. The voltage per cell is low, around 0.6V, and the available current depends on the intensity of the light. Efficiency is low, usually less than 5%, and the application is mainly to spacecraft and satellites which can obtain an unrestricted supply of sunlight at a very high intensity. A few remote systems in sunny areas can also make use of solar cells, but there is a major problem involved in keeping transparent surfaces clean enough to maintain enough light at the surface of the cells.

The original photo-voltaic cells used vacuum techniques, with a cathode of caesium metal and a nickel anode, producing a small current at a voltage of about 0.3V when illuminated. One of the main problems of any photocell generating system is the low density of energy in light, so that acres of cells are needed to produce the energy output of a small petrol generator. Another problem is that the energy content of light depends on its colour, with violet and ultra-violet the most energetic and red and infrared the least. Photocells for generating purposes must be able to release electrons under the lower-energy colours of light, requiring the use of materials which liberate electrons very easily.

Thermal generators make use of temperature differences to generate EMF. The use of thermocouples, which consist of junctions between two different metals (Figure 6.8), has been known for a long time. Junctions must exist in pairs, with one junction hot and the other cold, and the EMF is proportional to the temperature difference over a range of temperatures. The EMF

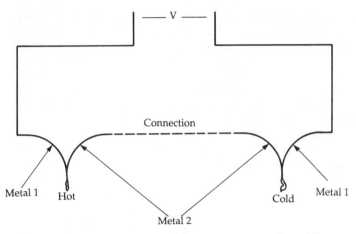

Figure 6.8 Thermocouple principles, using two junctions where different metals meet. The EMF is of the order of millivolts only.

reaches a peak at one value of temperature difference and then declines, so that there is an optimum temperature difference for any particular pair of metals used for a thermocouple. Because the principle has been known for over a hundred years, thermal converters have been available for most of this century, though they have never been common. One unit, made by Milnes in Tayport, Fife, Scotland, was widely sold for powering radio receivers in the later 1940's when a substantial number of homes had gas supplies but no mains electrical supply.

The EMF of each thermocouple is very small, a few millivolts at the most, so that a huge number of individual junctions must be connected in series in order to obtain a reasonable output. The current can be fairly high, however, limited mainly by the resistance of the metal junctions. The choice of metals affects the EMF and resistance values considerably, and the best combinations for the purposes of temperature measurement (like copper/constantan) are not necessarily ideal for power generation. Thermal generators have also been made using the principle of the thermionic valve, in which a high temperature is applied to a cathode which then releases electrons into a vacuum to land on an anode. The voltage and current levels of such devices are very low, however, and the method is not very practical for generating purposes.

Last on the exotic list is magnetohydrodynamic generation. The principle is of blasting a stream of hot gas through a magnetic field, and picking up the generated EMF on two opposite electrodes. The principle harks back to Faraday's principles of obtaining an EMF by moving a conductor in a magnetic field, and the main problem is to make the hot gas sufficiently conducting to achieve reasonable currents. The system has been used experimentally as a way of obtaining power from gas turbines without using rotating generators, or for making use of hot waste gases, but the need to make the gas conductive by such methods as seeding with sodium ions makes the scheme difficult to achieve.

Index

Index

Index